Fire and Rescue Service

Operational guidance
Railway Incidents

information & publishing solutions

Published by TSO (The Stationery Office) and available from:

Online
www.tsoshop.co.uk

Mail, Telephone, Fax & E-mail
TSO
PO Box 29, Norwich, NR3 1GN
Telephone orders/General enquiries: 0870 600 5522
Fax orders: 0870 600 5533
E-mail: customer.services@tso.co.uk
Textphone 0870 240 3701

TSO@Blackwell and other Accredited Agents

Contents

Foreword

Major incidents involving Railways in the United Kingdom are rare. Such incidents place significant demands on local Fire and Rescue Services and often require resources and support from other Fire and Rescue Services and emergency responders. However smaller scale incidents involving railways are more prevalent and these may require a response from any Fire and Rescue Service in England.

The Fire and Rescue Service Operational Guidance – Railway Incidents provides robust yet flexible guidance that can be adapted to the nature, scale and requirements of the incident.

The Chief Fire and Rescue Adviser is grateful for the assistance in the development in this guidance from a wide range of sources, including the Fire and Rescue Service, rail operators and rail industry experts.

It is anticipated that this guidance will promote common principles, practices and procedures that will support the Fire and Rescue Service to resolve incidents in this type of structures safely and efficiently.

The objective of the Fire and Rescue Service Operational Guidance – Railway Incidents is to provide a consistency of approach that forms the basis for common operational practices, supporting interoperability between Fire and Rescue Services, other emergency responders, railway and train operators and the rail industry. These common principles, practices and procedures are intended to support the development of safe systems of work on the incident ground and to enhance national resilience.

Operational guidance issued by the Department of Communities and Local Government promotes and develops good practice within the Fire and Rescue Service and is offered as a current industry standard. It is envisaged that this will help establish high standards of efficiency and safety in the interests of employers, employees and the general public.

This Guidance, which is compiled using the best sources of information known at the date of issue, is intended for use by competent persons. The application of the guidance does not remove the need for appropriate technical and managerial judgement in practical situations with due regard to local circumstances, nor does it confer any immunity or exemption from relevant legal requirements, including by-laws. Those investigating compliance with the law may refer to this guidance as illustrating an industry standard.

It is a matter for each individual Fire and Rescue Service whether to adopt and follow this Operational Guidance. The onus of responsibility for application of guidance lies with the user. Department for Communities and Local Government accept no legal liability or responsibility whatsoever, howsoever arising, for the consequences of the use or misuse of the guidance.

Section 3

Introduction

Purpose

3.1 This Operational Guidance is set out in the form of a procedural and technical framework. Fire and Rescue Services should consider it when developing or reviewing their policy and procedures to safely and efficiently resolve emergency incidents involving any aspect of Fire and Rescue Service operations involving railways.

3.2 The term 'rail infrastructure' is a general term encompassing:

- rail vehicles

- traction systems

- all aspects of the built rail environment including:
  - tracks
  - stations
  - termini
  - bridges
  - viaducts etc.

3.3 For the purposes of this guidance a rail system is defined as:

'Transport infrastructure managed for the mass transportation of people or goods, guided by one or more fixed rails.

This description is intended to include national rail, metro, tram and heritage rail networks. This may also include temporary rail systems. It will also be useful when dealing with incidents on rail systems associated with dockyards, nuclear installations, quarries or other large industrial undertakings.'

3.4 Although non Fire and Rescue Service organisations and agencies may use other more specific definitions for their own requirements, the above is the one most appropriate for Fire and Rescue Services to base their risk assessments and planning assumptions on.

3.5 A Fire and Rescue Service may respond to a wide range of incidents involving tunnels and underground structures that have the potential to cause harm and disruption to firefighters and the community.

3.6 The purpose of this guidance is to assist emergency responders to make safe, risk assessed, efficient and proportionate responses when attending and dealing with operational incidents involving the rail infrastructure.

3.7 Whilst this guidance may be of use to a number of other agencies, it is designed to provide relevant information for the Fire and Rescue Service in England relating to planning and operations for incidents affecting the rail infrastructure.

Scope

3.8   This guidance covers a wide range of incident types associated with the rail infrastructure that are likely to be encountered. It is applicable to any event regardless of scale, from incidents, such as small fires occurring on rural rail embankments to large scale collisions involving large numbers of Fire and Rescue Service resources and members of the public.

3.9   It focuses on the tactical and technical aspects of rail incidents to assist Fire and Rescue Services with:

- the development of safe systems of work

- interoperability at large or cross border incidents where more than one Fire and Rescue Service is in attendance

- multi-agency working.

3.10   This guidance covers the time period from the receipt of the first emergency call to the closure of the incident by the Fire and Rescue Service Incident Commander.

3.11   In addition to detailed tactical and technical information it also outlines the key operational and strategic responsibilities and considerations that need to be taken into account to enable the Fire and Rescue Service to train for, test intervention strategies and plan to ensure effective response to any incident involving the rail infrastructure.

Structure

3.12   This guidance is based on nationally accepted good practice. It is written as an enabling guide based around risk-critical operational principles rather than a strict set of rules and procedures. This is done to recognise local differences across England and elsewhere in the UK in terms of risk profiles and levels of resource.

3.13   Section 8 contains the main body of the guidance and is divided into three parts:

- Part A – Pre-planning considerations

- Part B – Operational considerations

- Part C – Technical considerations.

Part A – Pre-planning

Information supporting Fire and Rescue Service personnel when undertaking preparatory work for dealing with incidents that may occur in their service area involving railways.

This section covers planning considerations at both the strategic level when planning for service wide response options and for those associated with local site specific risks.

Part B: Fire and Rescue Service operational considerations

Guidance to Fire and Rescue Service staff on responding to and resolving railway incidents. It is structured around six emergency response phases common to all operational incidents.

The procedure detailed uses the incident command system decision making model as its foundation. It is a generic standard operating procedure for dealing with railway incidents that Fire and Rescue Services can adopt or adapt depending on their individual risk assessments and resources.

Each section of the generic standard operating procedure details extensive lists divided into:

• possible actions

• operational considerations.

It should be stressed that these are **not mandatory procedures.** They are a 'tool box' of operational considerations which will act as an enabling guide when dealing with railway incidents.

The generic standard operating procedure reflects the hazards and control measures of the national generic risk assessments relevant to railway incidents.

Part C – Technical considerations

Contains technical information and operational considerations that may be required by Fire and Rescue Service personnel for planning training and operations. It also references more detailed guidance that may be of interest to Fire and Rescue Services.

This part only contains information with an operational connotation and is not intended to be an exhaustive technical reference document.

Section 4

Legal framework

Introduction

4.1 Fire and Rescue Authorities need to be aware of the following legislation. It is relevant to command and control at operational incidents and also in the training environment.

4.2 This section does not contain detailed legal advice about the legislation. It is just a summary of the relevant legislation, as applied to Fire and Rescue Authorities firefighting procedures. You should confirm with your legal team on your Fire and Rescue Authority compliance with this legislation.

4.3 When considering this framework it is essential to recognise that any definitive interpretation of the legal roles and responsibilities imposed by legislation can only be given by a court of law.

4.4 For a full understanding of the responsibilities imposed by the legislation, and by the Fire and Rescue Service National Framework, reference should be made to the relevant legislation or the National Framework. It is also recognised that the range of legislation and guidance that could impact on the operational responsibilities of the Fire and Rescue Authority is extensive and each Authority should seek guidance from their own legal advisors.

4.5 The adoption of the principles set out in this guidance will assist Fire and Rescue Authorities in achieving suitable and sufficient risk assessments and appropriate corresponding risk control measures such as those referred to in this and other similar documents.

Primary Fire and Rescue Service legislation

FIRE AND RESCUE SERVICES ACT 2004

4.6 This is the main Act affecting Fire and Rescue Authorities. Amongst other things, it obliges Fire and Rescue Authorities (in section 7) to secure the provision of the personnel, services and equipment that efficiently meet all normal requirements and also to secure the provision of training for such personnel in relation to firefighting.

FIRE AND RESCUE SERVICES (EMERGENCIES) (ENGLAND) ORDER 2007

4.7 The Order obliges Fire and Rescue Authorities to make provision for decontaminating people following the release of chemical, biological, radiological or nuclear contaminants (article 2). It also requires them to make provision for freeing people from collapsed structures and non-road transport wreckages (article 3). The Order also obliges Fire and Rescue Authorities to use their specialist chemical, biological, radiological or nuclear or urban search and rescue resources outside their own areas to an extent reasonable for dealing with a chemical, biological, radiological or nuclear or urban search and rescue emergency (regulation 5).

CIVIL CONTINGENCIES ACT 2004

4.8   Section 2(1) states, among other things, that Fire and Rescue Authorities shall maintain plans for the purpose of ensuring that if an emergency occurs or is likely to occur the Fire and Rescue Authority is able to perform its functions so far as necessary or desirable for the purpose of preventing the emergency, reducing controlling or mitigating its effects or taking other action in connection with it.

THE CIVIL CONTINGENCIES ACT 2004 (CONTINGENCY PLANNING) REGULATIONS 2005

4.9   Fire and Rescue Authorities must cooperate with each other in connection with the performance of their duties under section 2(1) of the Civil Contingencies Act 2004. In addition, the regulations state that Fire and Rescue Authorities may facilitate cooperation by entering into protocols with each other (regulation 7), that Fire and Rescue Authorities may perform duties under section 2(1) jointly with one another and make arrangements with one another for the performance of that duty (regulation 8).

4.10   The Civil Contingencies Act 2004 (Contingency Planning) Regulations 2005 set out clear responsibilities for category 1 and category 2 responders and their need to participate in local resilience forums.

THE REGULATORY REFORM (FIRE SAFETY) ORDER 2005 (S.I. 2005/1541)

4.11   This legislation applies to all parts of the infrastructure with the exception of rail vehicles.

Primary health and safety at work legislation

HEALTH AND SAFETY AT WORK ETC ACT 1974

4.12   This Act applies to all employers in relation to health and safety. It is a wide ranging piece of legislation but in very general terms, imposes the general duty on Fire and Rescue Authorities to ensure, so far as is reasonably practical, the health, safety and welfare at work of all of their employees (section 2(1)).

CORPORATE MANSLAUGHTER AND CORPORATE HOMICIDE ACT 2007

4.13   Fire and Rescue Authorities will be criminally liable for the death of a employee if the way in which they manage or organise themselves:

- amounts to a gross breach of the duty of care owed to employees, and

- the gross breach causes an employees death.

4.14   Any alleged breaches of this act will be investigated by the police. Prosecution decisions will be made by the Crown Prosecution Service (England and Wales), the Crown Office and Procurator Fiscal Service (Scotland) and the Director of Public Prosecutions (Northern Ireland).

4.15   These regulations require Fire and Rescue Authorities, among other things, to make suitable and sufficient assessment of the risks to the health and safety of firefighters which they are exposed while on duty (regulation 3(1)(a)); to implement any preventive and protective measures on the basis of the principles specified in the regulations (regulation 4); to make arrangements for the effective planning, organisation, control, monitoring and review of the preventive and protective measures (regulation 5) and to provide such health surveillance as is appropriate having regard to the risks to health and safety which are identified by the risk assessment (regulation 6).

4.16   Safety Representatives and Safety Committees Regulations 1977 (as amended) and Codes of Practice provide a legal framework for employers and trade unions to reach agreement on arrangements for health and safety representatives and health and safety committees to operate in their workplace.

4.17   Health and Safety (Consultation with Employees) Regulations 1996 (as amended), sets out the legal framework which will apply if employers have employees who are not covered by representatives appointed by recognised trade unions.

Provision and Use of Work Equipment Regulations 1998

4.18   These regulations require Fire and Rescue Authorities to ensure that work equipment is constructed or adapted as to be suitable for the purpose for which it is used or provided (regulation 4(1)). Fire and Rescue Authorities must have regard to the working conditions and to the risks to the health and safety of firefighters which exist in the premises in which the equipment is to be used and any additional risk posed by the use of that equipment (regulation 4(2)). The regulations also contain provisions relating to maintenance, inspection, specific risks, information and instructions and training regarding work equipment.

Personal Protective Equipment at Work Regulations 1992

4.19   These regulations require Fire and Rescue Authorities to ensure that suitable personal protective equipment is provided to firefighters (regulation 4(1)). The regulations contain provisions in respect of the suitability, compatibility, assessment, maintenance, replacement, storage, information, instruction and training and the use of personal protective equipment.

4.20   Any personal protective equipment purchased by a Fire and Rescue Authority must comply with the *Personal Protective Equipment Regulations 2002* and be "CE" marked by the manufacturer to show that it satisfies certain essential safety requirements and, in some cases, has been tested and certified by an approved body.

Control of Substances Hazardous to Health Regulations 2002

4.21   Fire and Rescue Authorities must ensure that the exposure of firefighters to substances hazardous to health is either prevented or, where prevention is not reasonably practicable, adequately controlled (regulation 7(1)). Where it is not reasonably practicable for Fire and Rescue Authorities to prevent the hazardous exposure of firefighters, Fire and Rescue Authorities must, among other things, provide firefighters with suitable respiratory protective equipment (which must comply with the Personal Protective Equipment Regulations 2002 and other standards set by the Health and Safety Executive.

Dangerous Substances and Explosive Atmospheres Regulations 2002

4.22   Fire and Rescue Authorities are obliged to eliminate or reduce risks to safety from fire, explosion or other events arising from the hazardous properties of a "dangerous substance". Fire and Rescue Authorities are obliged to carry out a suitable and sufficient assessment of the risks to firefighters where a dangerous substance is or may be present (regulation 5). Fire and Rescue Authorities are required to eliminate or reduce risk so far as is reasonably practicable. Where risk is not eliminated, Fire and Rescue Authorities are required so far as is reasonably practicable and consistent with the risk assessment, to apply measures to control risks and mitigate any detrimental effects (regulation 6(3)). This includes the provision of suitable Personal Protective Equipment (regulation 6(5)(f)).

Confined Spaces Regulations 1997

4.23   A firefighter must NOT enter a confined space to carry out work for any purpose unless it is not reasonably practicable to achieve that purpose without such entry (regulation 4(1)). If entry to a confined space is unavoidable, firefighters must follow a safe system of work (including use of breathing apparatus) (regulation 4(2)) and put in place adequate emergency arrangements before the work starts (regulation 5).

The Work at Height Regulation 2005 (as amended)

4.24   This regulation replaces all of the earlier regulations in relation to working at height. The Work at Height Regulations 2005 consolidates previous legislation on working at height and implements European Council Directive 2001/45/EC concerning minimum safety and health requirements for the use of equipment for work at height (The Temporary Work at Height Directive).

Reporting of Injuries, Diseases and Dangerous Occurrences Regulations 1995

4.25   For the purpose of this section, regulation 3 is particularly relevant because it requires Fire and Rescue Authorities to notify the Health and Safety Executive of any "dangerous occurrences". Some examples of dangerous occurrences as defined in Reporting of Injuries, Diseases and Dangerous Occurrences Regulations 1995 relevant to Fire and Rescue Services operations at railway incidents include: 'any unintentional incident in which plant or equipment either (a) comes into contact with an un-insulated overhead electric line in which the voltage exceeds 200 volts; or (b) causes an electrical discharge from such an electric line by coming into close proximity to it'.

Specific legislation

Railways and Other Guided Transport Systems (ROGS) Regulations 2006

4.26   These regulations require that a Transport Undertaking or Infrastructure Manager may only operate if they have a Safety Management System in place which conforms to the requirements of the regulations. This specifically includes, *"provision of plans for action, alerts and information in the case of an emergency which are to be agreed with any public body, including the emergency services, that may be involved in such an emergency"*

The Fire Precautions (Sub-surface Railway stations) (England) Regulations 2009 Technical specification for interoperability – safety in railway tunnels

4.27   These regulations provide railway companies and enforcing authorities with a simple, clear regulatory framework within which to operate and address fire safety issues on sub-surface railway stations.

The Carriage of Dangerous Goods and Use of Transportable Pressure Equipment Regulations 2009

4.28   These regulations cover the carriage of radioactive materials by road and rail, and set out the requirements for placarding and markings.

Further reading

4.29   Operational guidance on the management of risk in the operational environment has been issued in the past. In particular, refer to:

- Fire Service Manual volume 2 (3rd edition) Incident Command

- Integrated Risk Management Plan guidance notes

- A guide to Operational Risk Assessment

- Health and Safety Executive guidance booklet HSG53: Respiratory protective equipment at work: A practical guide

- Striking the balance between operational and health and safety duties in the Fire and Rescue Service.

4.30   The adoption of the principles set out in this guidance will assist Fire and Rescue Authorities in achieving suitable and sufficient risk assessments and appropriate corresponding risk control measures such as those referred to in this and other similar documents.

4.31   The Fire Service College maintain a bibliography of technical guidance to which Fire and Rescue Services can refer (Fire service manuals, Fire Service circulars, Dear Chief Officer letters, technical bulletins, British and European Standards, Approved Codes of Practice, Health and Safety Executive guidance). In addition, technical guidance is available on the Department for Communities and Local Government website.

Fire and Rescue Service Operational Guidance – Railway Incidents

Section 5

Strategic role of operational guidance

Fire and Rescue Service Operational Guidance – Railway Incidents

Strategic perspective

5.1 Fire and Rescue Authorities and strategic managers with the Fire and Rescue Service are responsible for ensuring their organisation and staff operate safely when dealing with incidents involving railways. Their legal duties and responsibilities are contained in Section 4 of this guidance.

5.2 Fire and Rescue Services should continually assess the risk, in terms of the foreseeable likelihood and severity, of railway incidents occurring within their areas. This assessment should form part of their integrated risk management plan. The findings will help them ensure they have appropriate organisation, policy and procedures in place for dealing with railway incidents.

5.3 How do strategic managers know if they are providing, at least, the minimum level of acceptable service or possibly meeting their 'duty of care'? The following principles may assist Strategic Managers when determining the level of acceptable service and whether they are meeting their duty of care:

- operations must be legal and within the requirements of regulations

- actions and decisions should be consistent with voluntary consensus standards, and nationally recommended practices and procedures

- actions and decisions to control a problem should have a technical foundation and be based on an appropriate risk assessment

- actions and decisions must be ethical.

At the incident

5.4 'Response' can be defined as the actions taken to deal with the immediate effects of an emergency. It encompasses the resources and effort to deal not only with the direct effects of the emergency itself (e.g. fighting fires, rescuing individuals) but also the indirect effects (e.g. disruption, media interest). The duration of the response phase will be proportionate to the scale and complexity of the incident.

5.5 The generic key roles of the Fire and Rescue Services at rail incidents are:

- save life and carry out rescues

- fight and prevent fires

- manage hazardous materials and protect the environment

- mitigate the effects of the incident

- ensure the health and safety of fire service personnel, other category 1 & 2 responders and the public

- safety management within the inner cordon.

5.6   When responding to incidents involving railways the Fire and Rescue Service has strategic multi-agency responsibilities. These are additional, and in the main complimentary, to the specific fire and rescue functions that the Fire and Rescue Service performs at the scene. The strategic objective is to co-ordinate effective multi-agency activity in order to:

- preserve and protect lives

- mitigate and minimise the impact of an incident

- inform the public and maintain public confidence

- prevent, deter and detect crime

- assist an early return to normality (or as near to it as can be reasonably achieved).

5.7   Other important common strategic objectives flowing from these responsibilities are to:

- participate in judicial, public, technical or other inquiries

- evaluate the response and identify lessons to be learnt

- participate in the restoration and recovery phases of a major incident.

Values

5.8   The Fire and Rescue Service expresses its values and vision of leadership in the form of a simple model. The model has been named Aspire and is fully described in the *Fire and Rescue Manual (volume 2 Operations) – Incident Command*. It has at its heart, the core values of the service; which are:

- diversity

- our people

- improvement

- service to the community.

5.9   These values are intrinsic to everything Fire and Rescue Services strive to achieve at an operational incident, where they routinely serve all communities equally and professionally, with the safety and well being of their crews at the forefront of their procedures and reflecting on how well they performed in order to be better next time. It is important that core values are recognised and promoted by all strategic managers and fire and rescue authority members.

5.10   This guidance has been drafted to ensure that equality and diversity issues are considered and developed and has undergone full equality impact assessment in line with priority one of the Equality and Diversity Strategy.

Operational guidance review protocols

5.11   This operational guidance will be reviewed for its currency and accuracy three years from date of publication. The Operational Guidance Programme Board will be responsible for commissioning the review and any decision for revision or amendment.

5.12   The Operational Guidance Programme Board may decide that a full or partial review is required within this period.

Section 6
Generic Risk Assessment

Introduction

6.1   Due to the size and nature of the Fire and Rescue Service and the wide range of activities in which it becomes involved, there is the potential for the risk assessment process to become a time consuming activity. To minimise this and avoid having inconsistencies of approach and outcome, the Department for Communities and Local Government have produced a series of generic risk assessments. These generic risk assessments have been produced as a tool to assist Fire and Rescue Services in drawing up their own assessments to meet the requirements of the Management of Health and Safety at Work Regulations 1999.

6.2   There are occasions when the risks and hazards sited in any of the generic risk assessments may be applicable to incidents in tunnels and underground structures. However there are specific generic risk assessments that Fire and Rescue Services should consider when developing their policy and procedures for dealing with railway incidents. They have been used as the foundations of the information and guidance contained in this operational guidance.

6.3   Generic risk assessments of particular relevance to railway incidents

- 1.1 Emergency response and arrival at scene

- 2.7 Rescues from tunnels

- 4.2 Incidents involving transport systems –
  Rail http://www.communities.gov.uk/documents/fire/pdf/1829947.pdf

6.4   Fire and Rescue Services should use these generic risk assessments as part of their own risk assessment strategy not as an alternative or substitute for it. They are designed to help a Fire and Rescue Service to make a suitable and sufficient assessment of risks as part of the normal planning process. It is suggested that Fire and Rescue Services:

- Check the validity of the information contained in the generic risk assessment practices and identify any additional or alternative hazards, risks and control measures

- Evaluate the severity and likelihood of hazards causing harm, and the effectiveness of current controls, for example, operational procedures, training and personal protective equipment methodology

- Consider other regulatory requirements as outlined in Section 4 of this guidance

- Identify additional measures which will be needed to reduce the risk, so far as is reasonably practicable

- Put those additional measures and arrangements in place

- Fire and Rescue Services must plan the type, weight and speed of response to be provided to a particular location on the basis of reasonably foreseeable incident scenario. This should be decided on the basis of experience and professional judgement.

6.5 Once a suitable and sufficient assessment of the risks has been made, any additional measures and arrangements put in place have to be reviewed as part of the HSG 65 guidance.

6.6 To ensure the risk assessment remains suitable and sufficient Fire and Rescue Services should review the assessment to take into account, for example the learning outcomes from operational incidents, accidents etc.

6.7 Generic risk assessments provide a guide to the type of information, arrangements and training that should be given to the incident commander, firefighters and any other personnel likely to be affected.

6.8 Full guidance on the generic risk assessment is contained in Occupational health, safety and welfare: Guidance for Fire and Rescue Services: Generic Risk Assessment – Introduction.

Section 7

Key principles

Introduction

7.1 This operational guidance offers generic guidance to assist Fire and Rescue Authorities in their preparation for dealing with railway incidents as defined in Generic Risk Assessment 4.2 (Incident involving transport systems: rail). It is essential to consider this guidance and the relevant generic risk assessments in conjunction with local integrated risk management plans and local risk information to develop generic service wide plans, along with site specific variations and adjustments where necessary.

7.2 When planning for incidents involving railways Fire and Rescue Services should be aware that these can span administrative and governmental boundaries and therefore need to consider the involvement of a range of stakeholders including any Fire and Rescue Service affected.

7.3 To enhance the effectiveness of the local Fire and Rescue Service and site specific plans the Fire and Rescue Service should ensure that suitable and sufficient training and familiarisation is regularly undertaken to embed understanding of local risks and intervention strategies.

7.4 When attending railway incidents, Incident Commanders must determine and establish proportionate control measures over rail vehicle movements and traction current that take into account local standard operating procedures and relevant national guidance.

7.5 Where it is necessary for the implementation of control measures, such as, stopping rail vehicle movements and traction current being switched off, it is recommended that confirmation of implementation is received from the rail infrastructure manager before committing crews.

7.6 The rail industry has agreed not to unreasonably delay the implementation of proportionate control measures requested by the Fire and Rescue Services. However it must be recognised that some delay may occur for public safety purposes. Issues relating to public safety may be remote from any incident.

7.7 In some, extremely rare, circumstances the need for immediate action may be such that it may not be possible for Incident Commanders to await confirmation of implementation of control measures prior to committing crews. Such circumstances can include incidents where a delay in intervention could result in a saveable life being lost or preventing catastrophic escalation of the incident.

7.8 Where it is necessary for operational crews to work on or near the railway, Incident Commanders must ensure that appropriate safety officers are appointed and that they are adequately briefed.

7.9 Due to the complex and specialised nature of railway incidents, effective liaison at an early stage is essential. Incident Commanders must ensure that timely and appropriate liaison is established with the 'Responsible Person at Silver' or in their absence with Rail Control via Fire and Rescue Services Control.

7.10 Resolution of railway incidents is usually dependent upon the interoperability between a number of emergency responders and agencies. It is therefore essential that Incident Commanders identify all relevant agencies and duty holders and establish appropriate communications at an early stage.

7.11 When developing tactical plans for dealing with railway incidents, Incident Commanders will need to use knowledge of pre-planned intervention strategies and take into account all aspects of the circumstances of the incident (e.g.; rolling stock contents, topography, and other features) to ensure that firefighting, and rescue techniques and tactics are appropriate.

7.12 A significant feature of Fire and Rescue Services operations at railway incidents is access, egress and evacuation of the public. Incident Commanders should gather sufficient information to facilitate identification of an incident's location and appropriate access point to the infrastructure.

7.13 Railway incidents are often by nature linear, with limited access points. This can have a significant effect on the provision of equipment and personnel to scenes of operation. Incident Commanders should therefore carefully consider the affects of the geography of any incident on logistics, supply chains and crew welfare.

7.14 The ability of Fire and Rescue Service personnel to make and effective intervention is dependant on the severity of the incident, available systems and facilities, intervention strategies and the availability of resources and limitation of Fire and Rescue Services' equipment.

7.15 Railway incidents are often spread over large areas with command points remote from operations, Incident Commanders should therefore consider the early establishment of effective communications between the key points of the incident management structure.

7.16 The rail industry will normally undertake an investigation into the circumstances of any incident with any significant impact on safety or service delivery.

7.17 Fire and Rescue Services should ensure that the hand over of the scene is given to the British Transport Police/Home Office Police Service/ Rail Accident Investigation Branch, HM Inspector of Railways from the Office of Rail Regulation, or appropriate branch of the rail industry.

Section 8

Fire Service Operations

Fire and Rescue Service Operational Guidance – Railway Incidents

Part A
Pre-planning considerations

Strategic planning

General

8A.1   Pre-planning at a strategic level to ensure that Fire and Rescue Services develop and maintain an appropriate and proportionate response to railway incidents is fundamental to protecting the public, Fire and Rescue Service responders and mitigating the wider impact of any incident.

8A.2   Strategic planners will need to consider the relationship to the critical national infrastructure that is made by the national rail infrastructure in regard to:

- the national rail infrastructure interfaces with, and forms a significant part of, the wider national transport networks, with national and international passenger and commercial implications

- the topography and national geographical coverage of the national rail infrastructure is such that it provides a conduit for cabling for the National Grid and communication systems and other utilities.

8A.3   The national rail infrastructure in this country consists of a range of rail systems that in many instances are integrated:

- international rail systems, including freight and passenger services

- national commuter and freight services

- local tram systems

- metro rail systems

- local heritage railways

- industrial railways which may be permanent or temporary.

8A.4   The rail system's management of their own operations will usually be composed of:

- **Infrastructure managers**
  will maintain and control the rail system and vehicle movements, including many large city terminus stations and maintenance depots.

- **Train operating companies**
  operate the rail vehicles and are responsible for any passengers.

- **Freight operating companies**
  transport freight on the rail infrastructure and may also manage their own infrastructure in some areas.

- These roles may be performed by more than one company

> **Note:**
> More than one infrastructure manager and/or companies may be operating within a geographical area).

Strategic planning considerations and duties

8A.5 Planners should recognise that due to the complex and integrated nature of the national rail infrastructure, incidents may result in other related emergencies remote from the original incident. Even relatively minor incidents on a rail system have the potential for injury and significant disruption and loss, this could occur locally or over a wide area with potential national or international implications for commerce, tourism and travel.

8A.6 The potential for injury applies not only to Fire and Rescue Service personnel and other category 1 & 2 responders, but also to members of the public, including those who may be held on trains not directly involved in the incident, or otherwise remote, such as overcrowded platforms and stations.

8A.7 To ensure incident management is both effective and efficient Fire and Rescue Services must ensure that pre-planning is undertaken. This planning should ensure that dialogue takes place between the Fire and Rescue Service and the infrastructure manager/s for the rail systems to which they are likely to respond. This will involve establishing structures to ensure appropriate liaison at Bronze, Silver and Gold levels. (See 3.1).

8A.8 In compliance with this guidance Fire and Rescue Services must determine appropriate and proportionate responses and resources to rail incidents within their area. Additional considerations to inform this process may include:

- the complexity and relative importance of the rail infrastructure within its area
- integrated risk management plan response options
- discussion at regional resilience fora e.g. threat level
- the hazards associated with the individual rail system and the likely severity and impact of any incident
- information received from liaison with rail industry
- it will be necessary to ensure the suitability and sufficiency of the resources and response is tested.

8A.9 The selection and implementation of the most appropriate procedures will be based on an assessment of risk and the allocation of suitable and sufficient resources, to ensure incidents are resolved safely and effectively.

8A.10 The general duties of the Fire and Rescue Service in responding to emergency incidents are contained within the Fire and Rescue Services Act 2004. Further statutory obligations relating to rail incidents are contained in The Fire and Rescue Services (Emergencies)(England)Order 2007 http://www.legislation.gov.uk/uksi/2007/735/contents/made

8A.11   Responsibilities applicable to both category 1 and category 2 Civil Contingency Act responders and the direction to participate with local resilience fora are set out within the *Civil Contingency* Act *CCA (Contingency Planning) Regulations 2005.* http://www.legislation.gov.uk/uksi/2005/2042/contents/made

8A.12   Most infrastructure managers will have responsibilities under civil contingency legislation as 'category 2' responders to co-operate and share relevant information with the Fire and Rescue Service.

8A.13   The rail industry are valued partners, and as such should be incorporated in to fire, police and local authority emergency response plans. This will be helpful to Fire and Rescue Services when undertaking planning, training and operations.

8A.14   When developing Fire and Rescue Service response plans it is essential that planners are cognisant of existing regional multi-agency 'major incident procedures' and that these are complimentary.

Multi-level planning

8A.15   Planning for Fire and Rescue Service actions at railway incidents on each rail system may take place on three principal levels. An example of good practice is the planning arrangements for Network Rail (shown in table below).

Level	Description	Attendees
Gold (Level 1)	National Emergency Planning and Co-ordination Committee This is a planning committee comprising of strategic managers from category 1 & 2 responders, Government and the rail industry Due to the wide geographical spread of the Network Rail system this body covers all Fire and Rescue Service areas and therefore Fire and Rescue Service representation is made by Chief Fire and Rescue Advisor	Association of Chief Police Officers British Transport Police, Chief Fire Officers Association Chief Fire Rescue Adviser London Fire Brigade, London Underground Limited Network Rail HM Inspector of Railways from the Office of the Rail Regulation.

Level	Description	Attendees
Silver (Level 2)	Typically includes liaison meetings between rail industry and local Fire and Rescue Service to: • review recent operational incidents and identify learning opportunities • consider new developments and develop intervention and evacuation strategies • share relevant organisational policy information	Fire and Rescue Service Operational Policy and Regulatory Fire Safety representatives Rail industry policy representatives.
Bronze (Level 3)	Typically this will include: • Fire and Rescue Service Act 7(2)(d) and 9(3)d visits for gathering for operational pre-planning • training and liaison	Local Fire and Rescue Service responders and local rail managers.

Underpinning strategic knowledge

8A.16   The day to day management of a railway can appear complex. To inform and support strategic planning it is essential that Fire and Rescue Service personnel tasked with developing emergency response plans should have some underpinning knowledge in regards of the:

- geographical boundaries of different rail systems

- geographical boundaries of responsibility for infrastructure managers

- number and location of signal/control rooms for different lines controlled by the same infrastructure manager

- different types of traction power systems used

- different emergency procedures and rail staff responsibilities on different systems, and that these can be involved at the same incident

- differing responsibilities and authority of rail professionals on-site

- numbers of train operating companies running rail vehicles on the line

- nature and range of hazard associated with local rail systems

- Fire and Rescue Service intervention tactics.

8A.17 Liaison will assist in the development of a strategic operational response strategy and with developing and maintaining plans and procedures for effective and efficient operational response to any rail system to which the Fire and Rescue Service is likely to respond. This may include:

- combined intervention and evacuation strategies for rail systems

- agreed operational response to rail system with the neighbouring Fire and Rescue Service using an agreed intervention strategy

- local liaison structures with rail and other agencies

- information gathering methods for operational personnel in local planning and at the scene of an incident

- methods to exchange information for improving emergency response with rail network, emergency services and other agencies based on operational experience.

8A.18 Any strategy developed must provide safe systems of work to allow Fire and Rescue Service operations to commence before the arrival of the rail system's Responsible Person at Silver.

8A.19 A range of activities may be undertaken by the Fire and Rescue Service to examine the performance of plans following exercises or incidents with a significant impact. This may include the development of systems and processes to analyse and review the performance of plans, and/or liaison and exercise at all levels with:

- local resilience fora

- rail infrastructure managers

- the train operating companies

- category 1 responders

- other statutory agencies including category 2 responders

- freight operating companies

- the third sector agencies

- any relevant voluntary response organisation e.g., Radio Amateurs' Emergency Network (RAYNET) and WRVS, formerly the Women's Royal Voluntary Service.

Future developments

8A.20 Consultation between the rail industry and the Fire and Rescue Service will take place during the planning of any new rail systems. The rail industry may also consult for significant upgrading of existing infrastructure that may affect emergency evacuation or Fire and Rescue Service intervention.

8A.21 The benefits of early consultation ensure that:

- intervention and evacuation strategies are realistic and integrated
- facilities for fire and rescue purposes are suitable and sufficient
- the project can demonstrate meaningful consultation has taken place as part of planning of the development.

8A.22 As part of consultation on future developments including construction stages, it is important for Fire and Rescue Services to actively participate during the development of emergency planning assumptions and of an appropriate intervention strategy. This will ensure that the Fire and Rescue Service and infrastructure manager's expectations are realistic and reasonable. Additionally it is important to examine the consequences of intervention on the rail system's own evacuation procedures.

8A.23 To ensure a consistent approach to Fire and Rescue Service operations, it is recommended that intervention strategies, agreed with infrastructure managers, particularly in areas of complex or multiple rail systems maintain a commonality of approach in terms of the rail system's actions and responsibilities. This is particularly important for features such as:

- intervention systems
- facilities provided for Fire and Rescue Service response
- communications between infrastructure manager and Fire and Rescue Services
- liaison with Responsible Person at Silver.

Local planning responsibilities

8A.24 This section is intended to inform and advise at Silver and Bronze levels on the development of local plans. Some examples of where local plans would be required range from, the development of an intervention strategy for a new rail system within a Fire and Rescue Service area through to a local fire station's site specific predetermined on arrival tactics.

8A.25 The development of local plans should reflect Fire and Rescue Service policies, procedures and local risk assessments developed at the strategic level (see Chapter 1). Local plans should also reflect the guidance within this manual.

8A.26 Suitable arrangements should be put in place to gather relevant information to facilitate the development of local plans for all of the various rail systems and types of incidents that local Fire and Rescue Services may be called to attend.

8A.27 Plans should also be developed for any significant temporary local works and/or variations to existing plans affecting any rail systems, with consideration given to the affect of those changes on weight of attack, tactics and previous agreements.

Local planning liaison

8A.28   General liaison with stakeholders is essential to ensure that the necessary information is secured to inform plans for adequate, timely and effective response and to create a safe system of work when planning for and attending incidents.

Liaison with rail industry

8A.29   Industry specific information will be available from a variety of national and local sources, some examples include:

- Department for Transport Rail Group
- Passenger Transport Executive Group (covering the six Regional Executives)
- Network Rail
- Association of Train Operating Companies
- Freight Transport Association
- British Transport Police.

TRAM AND LIGHT RAIL OPERATORS:
- Birmingham to Wolverhampton – Midland Metro
- Birkenhead – Wirral Transport Museum
- Blackpool – Blackpool tramway
- London – Tramlink
- Manchester – Manchester Metrolink
- Nottingham – Nottingham Express Transit
- Seaton, Devon – preserved Seaton Tramway
- Sheffield – Sheffield Supertram
- Tyne and Wear – Tyne and Wear Metro.

8A.30   This is not intended to be an exhaustive list and there are many other operators of rail services.

Regional and inter-agency liaison

8A.31   National and local rail infrastructures do not necessarily recognise topographical or administrative boundaries and the potential for major incidents is not limited to any particular line or train operator. Rail incidents are likely to be complex and resource intensive and may, require responses from a number of emergency services, specialist teams and neighbouring Fire and Rescue Services. This should be taken into account and as a result close liaison between regional resilience partners is essential when pre-planning.

Local planning information

8A.32 Following relevant research, Fire and Rescue Services should ensure that detailed local plans are prepared to include some or all of the following information:

- Fire and Rescue Service Act 2004 7(2)(d) visits
- access
- rendezvous points
- premises information boxes
- station control rooms
- intervention points
- ventilation systems
- fixed installations
- communications
- traction current supply system
- hazardous materials
- line speeds
- complex locations.

Fire and Rescue Service Act 2004 7(2)(d) and 9(3)(d) Visits

8A.33 When arranging for visits to rail systems, officers should be mindful of the limitations that may apply to accessing the infrastructure. When developing detailed plans arrangements should be made to ensure visits are arranged to limit, as far as possible, the impact on the rail system's operations.

Access

8A.34 All practical and reasonable areas of access, on to the rail infrastructure these may include:

- stations (both surface and sub surface)
- tunnels
- intervention points/emergency response locations
- cuttings
- bridges
- level crossings
- sidings and depots

- gates and hard standing for appliances

- appropriate maintenance access points.

Rendezvous points

8A.35   When determining the most suitable position for rendezvous points, consideration must be given to:

- crew safety

- access for appliances

- effective communications

- plans boxes

- water supplies.

Infrastructure control rooms

8A.36   An understanding of the facilities afforded by infrastructure control rooms will assist in determining the means by which an incident can be managed, these may include:

- location

- alternative access/egress

- close circuit television

- public address systems.

> **Note:**
> There are variations in the provision and location of control rooms, full use of 7(2)(d) and 9 (3) (d) visits should be made to determine the presence and location of station control rooms.

Intervention points/emergency response locations

8A.37   These are locations that can be used for means of access for an emergency response. Emergency response locations will also provide integrated facilities for Fire and Rescue Service intervention and managed evacuation by the relevant infrastructure manager incorporating train design, cross passages and rail managed evacuation trains. They may also incorporate evacuation facilities for members of the public. They can vary greatly from basic access stairs to complex purpose built structures. Crews should be aware of the following features:

- location

- rendezvous points

- access arrangements
- plans
- water supplies
- communication facilities.

Ventilation systems

8A.38 Some sub-surface stations and rail tunnels now have ventilation systems which may assist in the control of the fire/accident environment. Crews should be aware of the type, location, and operation of the control systems. Types of system are described in some detail in national guidance for dealing with incidents in tunnels and underground.

> **Note:**
> At incidents involving fire or hazardous materials ventilation systems should not be turned off or re-configured until a risk assessment has been made and the full consequences of these actions to the public, firefighters and any fire development are known.

Fixed installations

8A.39 Fixed installations to assist firefighting operations are provided in some locations throughout the rail infrastructure. The location, use, and implications of their operation should be known and understood. Fixed installations available may include:

- automatic fire detection systems
- sprinkler systems
- inert gas systems
- 110v electrical supplies for Fire and Rescue Service use
- fire mains/hydrants
- communications systems.

Traction power systems

8A.40 Rail vehicles use one or more of the following types of traction power; electricity, diesel, steam or battery. Identification of the power systems present during the planning stage will inform firefighting tactics and enhance firefighter safety.

Hazards to firefighters

8A.41   In general, the number of hazards facing firefighters and the likelihood of the associated risks occurring will vary in line with complexity in the rail infrastructure and its geographical location. This information should be considered in conjunction with the national generic risk assessment and technical information within this guidance.

Types of rail vehicles

8A.42   There are many different types of rail vehicles in use across the infrastructure with wide variations in physical dimensions, capabilities, construction materials, use and location of on board facilities eg; generators, data recorders etc. This degree of variation can significantly affect the risks to firefighters, and it is therefore imperative that plans include the risk critical aspects of vehicles that are likely to be encountered.

Complexity

8A.43   At many locations different rail systems may interface, these areas will be under the control of more than one infrastructure manager, often with a significant degree of complexity. When developing plans for these locations, Fire and Rescue Services should be aware that different infrastructure managers may be involved at the same incident. Plans should identify how the infrastructure managers for these locations work with others to secure firefighter safety.

Availability of local plans/training/review

Availability of plans

8A.44   Local plans should be readily available in appropriate formats to support the needs of first response crews, Incident Commanders, command support and elsewhere as required by local Fire and Rescue Service arrangements and the National Incident Command System. Where appropriate, consideration should also be given to sharing plans with other agencies and organisations.

Training and exercising

8A.45   Rail operators will periodically test emergency response arrangements. Where possible, local Fire and Rescue Service responders should participate in these exercises. The mutual benefits of joint training and exercising include:

- greater inter service understanding

- confidence and awareness of relevant roles and functions

- local familiarisation with key personnel and features of infrastructure.

8A.46   Local Fire and Rescue Services will wish to prepare and develop fire crews awareness of on arrival, intervention tactics and review of operations against the National Occupational Standards on Emergency Fire Service Management, through:

- on-site testing of Fire and Rescue Service aspects of plan

- full scale multi-agency exercises

- Fire and Rescue Service Act 2004 7(2)(d) and 9(3)(d) visits, internal lectures and presentations for Fire and Rescue Service responders.

Review

8.47   Large parts of the rail infrastructure are subject to significant on-going change and modification. Therefore Fire and Rescue Services should ensure that local plans are regularly reviewed and updated. This may be either periodically or at key milestones in the case of refurbishment/construction projects.

8.48   Following any significant railway incident/accident, a full and robust review of all Fire and Rescue Service policies, standard operating procedures and memorandum of understanding should be undertaken. Where these are amended or changed this should be communicated to relevant staff.

Part B

Operational considerations – Generic Standard Operating Procedure

Introduction

8B.1 It is useful to see the emergency incident response phases in the context of the typical stages of an incident as referred to in Volume 2 Fire Service Operations Incident Command Operation Guidance and the Fire Service Guide – Dynamic Management of Risk at Operational Incidents, this is shown below:

Stages of an Incident (Dynamic management of risk)	ICS Decision Making Model Links	Generic Standard Operating Procedure (GSOP) Response Phases
		1. Mobilising and en-route
Initial Stage	• Incident information • Resource information • Hazard and safety and information	2. Arriving and gathering information
Development Stage	• Think • Prioritise objectives • Plan	3. Planning the tactical plan
	• Communicate • Control	4. Implementing the tactical plan
	• Evaluate the outcome	5. Evaluating the tactical plan
Closing Stage		6. Closing the incident

8B.2 The Generic Standard Operating Procedure has been derived by breaking down an incident into six clearly identified phases which have been taken directly from the decision making model.

8B.3 The purpose of this section is to cover possible actions that may need to be undertaken at each of the six stages of the incident and then offer some possible considerations that the incident commander and other Fire and Rescue Service personnel may find useful in tackling the challenges and tasks that they are faced with.

8B.4 This Generic Standard Operating Procedure is not intended to cover every eventuality however it is a comprehensive document that can be used by planning teams, who need to write standard operating procedures, and responding personnel alike.

8B.5 Further detailed and technical information on specific rail incident related hazards are covered in Section 8 part C of this operational guidance.

8B.6   The decision making model comprises of two major components. These are the deciding and acting stages.

DECIDING	ACTING

8B.7   In seeking to resolve a railway incident an Incident Commander (IC) will use their knowledge and experience to identify the objectives to be achieved and formulate an appropriate tactical plan of action.

Emergency incident response phases

1	Mobilising and en-route
2	Arriving and gathering information
3	Formulating the tactical plan
4	Implementing the tactical plan
5	Evaluating the tactical plan
6	Closing the incident

Phase 1 – Mobilising and en-route

Phase 1 – Actions	
Mobilising and En-route	
1.1	Initial call handling
1.2	Assess the level and scale of the incident
1.3	Mobilise appropriate resources to the incident, marshalling areas and/or predetermined rendezvous points (RVPs)
1.4	Access incident specific information en-route
1.5	Notify relevant agencies

Considerations

1.1 Initial call handling

8B.8   As with any incident the handling of the initial call is of critical importance to ensure that the correct predetermined attendance (PDA) is mobilised. In handling the call the mobilising centre operator will need to gather as much information from the caller as possible. If there is any doubt as to the size and scale of the incident, the predetermined attendance should be scaled up rather than down.

1.2 Assess the level and scale of the incident

8B.9   Mobilised crews may contact infrastructure manager via Fire and Rescue Service Control whilst still en-route. This could provide additional information regarding the location, access locations and type of incident. This may be particularly useful in remote areas.

1.3 Mobilise appropriate resources

8B.10   Specific risks will attract a range of different mobilising solutions; these will normally be determined in the planning stage and may include variations in weight of attack, attendance to specific locations, dual attendances, specialist resources and advice.

8B.11   Fire and Rescue Service Controls should utilise any site specific plans to enhance mobilising information to crews, particularly when mobilising to complex locations within the rail infrastructure such as large termini or underground complexes.

1.4 Access incident specific information en-route

8B.12   A key aspect for dealing with incidents on the rail infrastructure is securing safe and effective access to the scene. This will often form part of predetermined intervention strategies for known locations, however for large parts of the infrastructure it is essential to narrow down the possible location so that appropriate points to enter the infrastructure can be identified.

- Mobilised crews should combine local knowledge with site incident specific information from on board systems whilst en route to identify: rendezvous points, predetermined on arrival tactics, points for initial information gathering on arrival

- Mobilised crews should use any available information to consider the likely risks, hazards and control measures they may face when mobilised to certain known parts of the rail infrastructure

- Mobilised Incident Commanders should use available information to select and implement safe systems of work when mobilised to certain known parts of the rail infrastructure including wind direction and topography.

1.5 Notify relevant agencies

8B.13   Fire and Rescue Service Controls should maintain contact details of infrastructure managers for rail networks that their Fire and Rescue Service may attend. This will allow information to be gathered in a timely manner and support the effective passage of information between the relevant parties.

8B.14   On most occasions infrastructure managers will be aware of incidents occurring on their infrastructure; however it is good practice for Fire and Rescue Service Controls to inform the relevant infrastructure manager of any incidents being attended by their Fire and Rescue Service.

8B.15   Resolving incidents on the rail infrastructure is often a result of a multi-agency effort. Fire and Rescue Service Controls should consider sharing relevant details about calls being attended by their Fire and Rescue Service with surrounding Fire and Rescue Services, other category 1 responders and attending agencies.

8B.16   Early dialogue between Fire and Rescue Service Controls and infrastructure managers will assist with the identification of:

- location and access

- incident type.

8B.17   This will be particularly useful where Fire and Rescue Service resources may be travelling long distances, or where the call is to complex areas of the infrastructure.

Phase 2 – Arriving and gathering information

Phase 2 – Actions Arriving and gathering information	• Incident information • Resource information • Hazard and Safety and information
2.1	Confirm location
2.2	Confirm incident type
2.3	Identify access routes
2.4	Identify type of infrastructure/rail vehicles involved
2.5	Confirm use of rail vehicles
2.6	Liaise with persons on scene
2.7	Use local knowledge
2.8	Identify available resources
2.9	Identify risks and hazards

Considerations

2.1 Confirm location

8B.18   Railways can be located in remote, rural or built up urban areas comprising of simple or complex infrastructure. To counter this and to assist with pre-planning and risk management on arrival, it is essential that Incident Commanders make every effort to identify the precise location and appropriate access points to the infrastructure.

8B.19   Incident Commanders may use a variety of information sources to inform this such as:

- local pre-planning information

- mobilising information

- information from infrastructure managers

- local risk information

- on-site information provided for Fire and Rescue Service use

- local knowledge of crews/topography.

8B.20 On arrival Incident Commanders may use a number of additional means to identify the location to Fire and rescue Service Control and/or supporting appliances such as:

- signal number

- bridge number

- overhead line support number

- quarter-mile post at the track side

- electrical substation name plate

- nearby station or level crossing

- significant geographical features

- point number

- local mapping systems (such as Ordinance Survey maps and 'six figure grid reference')

- nearest road or intersection.

8B.21 On determining the precise location of any incident, Incident Commanders must ensure that an appropriate message is sent to Fire and Rescue Service Control confirming location.

2.2 Confirm incident type

8B.22 There are various types of incident which fall into the above categories such as:

- fire on Infrastructure

- fire on rail vehicle

- derailment

- person(s)/vehicle trapped on rail infrastructure

- flooding

- hazardous materials

- collisions including with road vehicles or rail infrastructure ('bridge strikes')

- explosions

- terrorist related incident.

2.3 Identify access routes

8B.23 It is always preferable for Fire and Rescue Service crews to gain access to the infrastructure by means designed for public or Fire and Rescue Service access purposes. This principally involves:

- stations; or
- emergency response locations/ intervention points
- purpose built walkways.

8B.24 These locations would normally be preferred as a means of access. The principal advantage is that Fire and Rescue Service facilities are normally provided to assist operations and protect people from harm. Alternative access can be obtained by using facilities such as:

- use of rail vehicles, designed for carrying passengers (this does not include open flat bed rail wagons)
- access gates
- level crossings
- cutting through fences.

8B.25 The Incident Commanders will assess the urgency of the situation when determining the most appropriate method of accessing the infrastructure.

8B.26 Some Fire and Rescue Services have obtained specialist vehicles for use at rail incidents to mitigate access issues to specific infrastructure.

8B.27 Fire and Rescue Service personnel must not move from an area intended for normal public use (e.g. station platforms or the public highway) to an area *on or near the railway*, where there is a hazard from rail vehicles or the infrastructure, without first implementing appropriate control measures. Any signage provided should be considered as part of any risk assessment.

2.4 Identify type of infrastructure/rail vehicles involved

8B.28 Fire and Rescue Service crews should be aware of the types of infrastructure to which they are likely to respond to assist the Incident Commanders to determine:

- who is responsible for the management of the infrastructure (this may involve more than one infrastructure manager)
- what types of rail vehicle are likely to be encountered
- what traction systems are present
- features of the infrastructure that may represent additional risks to firefighters including:
  - bridges
  - tunnels

- viaducts

- embankments

- near open water

- utilities and electrical systems outside the control of the infrastructure manager such as National Grid supply cables

- cuttings.

- Facilities of the infrastructure that may be provided for Fire and Rescue Service use including:

  - emergency response location or intervention point

  - firefighting mains

  - hard standing

  - communications systems

  - on-site information

  - emergency lighting

  - electrical supplies for Fire and Rescue Service use

  - firefighting lobbies

  - ventilation systems.

2.5 Confirm use of rail vehicles

8B.29 Incident Commanders should gather relevant information regarding the specific use of any affected rail vehicles for example:

- whether affected rail vehicles are passenger or freight

- number, type and use of cars/carriages/units

- estimated number of passengers

- any evacuation information/strategies implemented

- any information regarding mobility impaired passengers

- type and quantity of freight carried

- information on hazardous materials (e.g.; via the national rail system's Total Operating Processing System).

2.6 Liaise with persons on scene

8B.30 Rail incidents can involve more than one Fire and Rescue Service, other emergency services, other agencies and commercial undertakings. Gathering and providing relevant timely information to and from interested parties is critical to successful outcomes.

8B.31 Incident Commanders should identify those representatives on-site and others that may be required and establish appropriate liaison structures, up to and including locally agreed arrangements for dealing with major incidents.

8B.32 At this stage of the incident liaison is likely to be limited to:

- local police service
- British Transport Police
- The Ambulance Service
- infrastructure managers
- train operating companies
- utility companies.

2.7 Use local knowledge

8B.33 Fire and Rescue Service crews and Incident Commanders should use knowledge of previously agreed railway intervention strategies and prior training in conjunction with any locally available plans and risk information to inform operational strategies and tactics.

2.8 Identify available resources

FIRE AND RESCUE SERVICE RESOURCES

8B.34 Fire and Rescue Services may have a range of specialist and non-specialist equipment and trained personnel to assist in dealing with various aspects of rail incidents, examples of which are:

a. detection identification monitoring equipment

b. enhanced command support

c. hazmat officers

d. high volume pumps

e. incident response units

f. inter-agency liaison officers

g. press liaison officers

h. thermal image cameras

i. short circuiting devices

j. specialist rail vehicles

k. urban search and rescue dog teams

l. urban search and rescue modules and teams.

8B.35 Incident Commanders will need to gather relevant information about Fire and Rescue Service resources in attendance and en-route and assess whether additional local or national resources are required.

NON FIRE AND RESCUE SERVICE RESOURCES

8B.36 Infrastructure managers will have access to equipment and personnel that may be available to support Fire and Rescue Services at any rail incident. This can be identified through local liaison and requested via agreed pre-planned procedures. This may include:

- telecommunications equipment
- lighting
- inflatable shelters
- video cameras and playback facilities
- airwave communications fitted to command vehicle's and helicopter
- heavy lifting equipment and cranes
- various specialist teams to assist in:
  - lifting and moving rail vehicles
  - body recovery
  - humanitarian support
  - rail cutting equipment
  - manually operated rail trolleys.

2.9 Identify risks and hazards

8B.37 This section should be read in conjunction with Generic Risk Assessment 4.2: Incidents involving transport systems – rail.

Phase 3 – Planning the response

Phase 3 – Actions Planning the response	• Think • Prioritise objectives • Plan
3.1	Identify and prioritise objectives
3.2	Establishing proportionate control over railway
3.3	Formulate and transmit appropriate messages
3.4	Choose appropriate access and egress routes
3.5	Select and establish relevant cordons
3.6	Select appoint and brief appropriate safety officers
3.7	Actions of deployed crews
3.8	Determine firefighting tactics
3.9	Carry out rescues
3.10	Resolve hazmat issues
3.11	Establish effective systems for liaison

Considerations

3.1 Identify and prioritise objectives

INCIDENT COMMAND

8B.38   Setting of objectives at rail incidents should be approached in the same way as all other operational incidents and should observe the principles established in the National Incident Command System.

COMPLEXITY

8B.39   Railways can be complex and are hazardous working environments, requiring a flexible approach to be adopted when planning a tactical response. Fire and Rescue Services, in partnership with the infrastructure managers, must co-operate and respond proportionately to any incident in order to serve the community, reduce risk to responders, and reduce the wider impact of the incident.

TRAINING

8B.40   Fire and Rescue Service Incident Commanders and LC crews should use their training and knowledge to identify the hazards, risks and control measures appropriate to the rail system they are responding to. (See Generic Risk Assessment 4.2: Incidents involving transport systems – rail).

INCIDENT TYPE

- **Rescue**

  Rescue operations at railway incidents could range from a single person trapped by a rail vehicle, to extremely complex and large scale operations, involving multiple rescues and casualties, undertaken over several days.

- **Fire**

  Fires could range from a small smouldering fire, through to rail vehicle carriages becoming fully involved, at inaccessible locations.

- **Hazmat**

  Incidents could range from leaking valves, through to significant spillages or ruptures of goods in transit. In addition to any hazardous materials that may be found on or in vehicles and infrastructure.

LEVEL OF CONTROL

8B.41   The first objective is to establish an environment for responders and casualties that proportionately protects against the hazards presented by the rail system and the incident to be confronted.

8B.42   Stopping trains and/or isolating traction current should only be requested in order to save life or property. Consideration should be given to running trains 'at caution'.

8B.43   Incident Commanders will need to determine the level of controls to be applied to the incident balanced against the potential risks caused to people stranded on trains, remote from the incident, as well as the associated disruption and cost to the rail industry. This assessment may include:

- passengers alighting from trains that have stopped outside stations and walking along tracks that are still live

- overcrowding of trains and platforms

- physical and mental distress of passengers held on trains potentially made worse by hot or cold conditions, or failure of air conditioning particularly when in tunnels

- disruption to trains over a widespread area.

8B.44   Intervention by the Fire and Rescue Service may lead to increased risks to firefighters, long delays, increased risk to passengers and the economic loss. Consequently, there may be circumstances when least risk is represented by allowing a small fire to burn itself out under monitored conditions from a safe distance.

3.2 Establishing proportionate control over railway

8B.45   When the Incident Commander has determined the incident objectives, it will be necessary to establish the level of control to be implemented over traction current and train movements as one of the first rail specific considerations. There are four levels of control that an incident commander may apply to control rail vehicle movements and traction current, as follows:

1. **Inform the infrastructure manager of an incident on or near the railway**

   These types of incident could include, for example, a small smouldering fire that the Incident Commander believes may be safely monitored until burnt out. Another example may be,a 'bridge strike' where a lorry has wedged under a rail bridge, but with no obvious damage to the rail infrastructure. The Incident Commander may wish for the infrastructure manager to send specialist rail personnel to inspect the rail lines, to see if any miss-alignment has occurred.

   A bridge strike where the Incident Commander informed the infrastructure manager of the nature of the incident.

2. **Request rail vehicles are 'run at caution'**

   This can be used when there is a need to slow train vehicle movements, by notifying drivers that there are people on or near the rail infrastructure. In these circumstances the driver will adjust their speed to ensure that the vehicle can be brought safely to a halt, if required. Some examples may include Fire and Rescue Service crews extinguishing trackside fires over 3m from live current. This would not be an appropriate control measure on some systems, using driverless vehicles.

   As cross winds could cause smoke to obscure the train drivers vision the Incident Commander requested trains run at caution.

3. **Request that rail vehicles are stopped**

   This will be used where there is a risk of people being injured by vehicle movements.

4.   It should be noted that this can take time to implement, as vehicles will have to reach or receive a 'stop' signal before the Fire and Rescue Service can be provided with a guarantee.

At this incident there was no rail electrical hazard, consequently the Incident Commander requested rail vehicles stopped.

5.   **Request power off**

This will be used when there is a significant risk of people or resources coming into contact with live electrical traction current. This will not necessarily stop all train movements, for example diesel vehicles will be unaffected, and high speed electric vehicles can coast for some distance.

Where Fire and Rescue Service operations need to take place within 3 metres of any traction current Incident Commanders may request electrical isolation of relevant track sections if appropriate.

If the incident involves overhead line equipment then there is a risk that residual current may remain, or near by high voltage power cables may induce an electrical charge into the overhead line equipment. Therefore when operating closer than 1 metre to overhead line equipment including operations such as ladder pitching or directing firefighting jets, Incident Commanders must make timely requests for isolation and earthing of relevant overhead line equipment sections to be carried out. This can only be undertaken by rail system personnel and confirmed by infrastructure managers.

The impact of power off can affect train vehicle movements over a very wide area, covering several counties.

The size and nature of the incident, combined with complexity of the rail system involved or adjacent systems will determine what level of control should be applied to the incident. More than one level of control could apply to an incident.

At larger incidents this may include arrangements for using rail vehicles to assist with the recovery stage.

It must be noted that these controls will only affect the movement of trains and the traction power supply. Other hazards, including for example, any third party high voltage electrical systems including the National Grid or infrastructure electrical systems, may remain live. These may be controlled using established procedures for securing electrical safety.

At locations where high voltage electrical supplies run adjacent to the area of operations there is a risk that an electrical current will be induced into any isolated cable.

It should be noted that more than one level of control may apply to single incident.

Once the Incident Commander has determined the extent and level of control required, a message should be formulated and transmitted. An exception to this is when the Fire and Rescue Service has agreed with the infrastructure manager that the on-site Responsible Person at Silver has the authority and ability to implement these controls locally. In these circumstances a message confirming local implementation should be transmitted to Fire and Rescue Service Control.

3.3 Formulate and transmit appropriate messages

8B.46 Messages to Fire and Rescue Service Control which may be subsequently relayed to the infrastructure manager require accuracy in formulation and transmission.

8B.47 To assist the infrastructure manager to implement safe systems of work, facilitate Fire and Rescue Services operations and reduce risk, it is necessary for the Fire and Rescue Service to include relevant details such as:

- if people are on or near the railway
- the location of the incident
- what level of control is required over which infrastructure
- over what geographical area controls should be applied
- nature of the Fire and Rescue Service activity being undertaken.

8B.48 Associated details from informative messages on the nature of the incident will also provide useful information to the infrastructure manager on potential rail system implications.

8B.49 It must be noted that it is the Fire and Rescue Service's expectation that no unreasonable delay should occur in the implementation of the request for appropriate and proportionate control over the rail system. The implementation of a Fire and Rescue Service request is not conditional on the Responsible Person at Silver being at the incident.

8B.50 The Incident Commander and Fire and Rescue Service Control should understand that the content and accuracy of the information exchanged will inform the urgency of the request, and define the areas that the request applies to. Following a request the rail system's control may, without disproportionate delay, take steps to protect passengers remote from the incident, being held in positions that may adversely affect wider passenger safety.

8B.51 At incidents involving complex rail infrastructure the affected rail system's control will contact any adjacent system to ensure Fire and Rescue Service safety.

8B.52 The infrastructure manager will confirm when the request has been implemented.

3.4 Choose appropriate access and egress routes

8B.53 When responding to an incident it is usually preferable to use the predetermined intervention strategy specific for the rail system or location. This will consist of the method of access and egress to the rail infrastructure and rail vehicles by the Fire and Rescue Service, as agreed with the infrastructure managers.

8B.54 Intervention strategies normally include locations designed or mutually agreed with the infrastructure managers for providing suitable Fire and Rescue Service access to the rail system and vehicles, for example:

- stations
- intervention point/evacuation point
- emergency response location
- steps
- gates
- level crossings
- tunnel portals
- hard standing locations.

8B.55 These locations will normally provide the initial rendezvous points/strategic holding areas. Such locations may also be provided with facilities to assist access to the scene and provide facilities for firefighting and rescue operations.

8B.56 Some incidents may involve difficult access to the rail system. In these cases it may be necessary to send part of any attendance to a vantage point for an assessment of the incident to be communicated to Fire and Rescue Service responders.

8B.57   It is recognised that in rare circumstances the situation may be so urgent that access to the system must be obtained by other means. This may mean cutting through rail fences or moving across other's property. In these circumstances the infrastructure managers must be informed of where the access is being made from in addition to an appropriate request for control over train movements and/or traction current.

8B.58   In all circumstances crews should receive an appropriate briefing, the details of which should include:

- the task to be performed
- the defined working area
- the path to and from the work area
- evacuation signals and corresponding actions.

8B.59   In all cases, when the Incident Commander is determining where and how to deploy crews and resources, additional risks posed by the type of incident such as hazardous materials should be considered as a contributing factor when determining safe systems of work.

8B.60   For incidents at locations with difficult access, arrangements may be put in place to use rail vehicles to gain access. These vehicles may consist of:

- the rail system's own vehicles
- specialist Fire and Rescue Service rail vehicles i.e. road/rail vehicles or trolleys.

8B.61   The use of the rail systems own vehicles will depend on factors such as:

- the type and severity of incident
- the hazards involved
- the infrastructure involved (for example a twin bore tunnel may protect from a fire in the other bore or a tunnel with fire ventilation may protect from fumes down stream)
- the rail systems intervention strategy and staff policy
- the availability and safety of suitable rail personnel.

8B.62   Some Fire and Rescue Services have obtained specialist rail vehicles for use at incidents. The benefits of using such vehicles are:

- reduce travel times
- reduce crew fatigue
- extended working durations
- assist with the recovery stage
- reduce the overall impact of the incident.

8B.63   Rail vehicles that have not been designed to carry passengers should not be used as a means of transporting crews.

3.5 Select and establish relevant cordons

8B.64   Rail incidents will often require the implementation and management of effective cordons in order to:

- protect people from rail hazards

- prevent a worsening of the incident

- preserve the scene for investigators

- define the safe working area

- assist with command and control.

8B.65   Certain features of the rail infrastructure can be used to good effect to define the extent of the Fire and Rescue Services' working area. This could include:

- stations

- platforms

- tunnel portals

- fences

- bridges

- signals

- tunnel cross passage doors or shafts

- marker posts.

8B.66   Within a cordon it may be appropriate to define paths for access and egress to the scene of operations. This will assist in reducing risks and preserving evidence. While any Fire and Rescue Service cordon is in place care should be taken to ensure access and egress is controlled and the number of personnel operating in the area is kept to a minimum.

8B.67   As soon as is reasonably practicable a reduction in cordon size should be implemented, in consultation with other agencies and the Responsible Person at Silver. Consideration should also be given to any reasonable requests for adjustment to Fire and Rescue Service operations in order to provide restoration of service to all or part of the rail system.

8B.68   It would be good practice for discussion to take place with the Responsible Person at Silver to help identify the rail specific hazards within the cordon, in order to assist the briefing of crews and partners.

8B.69   The cordon at a rail incident is likely to be linear in nature, and may cover a considerable distance. It may be necessary within or alongside the cordon to position staging posts, marshalling areas or command facilities. It would be beneficial to link cordons with sectorisation which is likely to be unique for this type of incident. Personnel accountability would increase in emphasis, as personnel would have a number of access/egress points, with co-operation between sectors/access points being extremely important.

3.6 Select, appoint and brief appropriate safety officers

8B.70   Any safety officers appointed should be briefed on the specific nature of the hazards they are responsible for monitoring, and the actions to be taken if additional control measures are required. Safety officers should also ensure that the position they take to monitor operations does not place them at risk from rail related hazards.

8B.71   This will be particularly the case for safety officers appointed to warn of rail vehicle movements. Safety officers should only be appointed for this role in situation when any person is at risk from rail vehicle movements. This will be implemented in rare circumstances when the Incident Commander believes there is a humanitarian imperative, where people are at risk from rail vehicles and that the appropriate control measures have not yet been confirmed as in place.

8B.72   Part of the assessment of whether to deploy safety officers to warn for rail vehicle movements and where they should be positioned to perform the duty should be subject to the following considerations:

- the availability of rail professionals to undertake the role of 'lookout'

- speed and stopping distances of rail vehicles (see below)

- distance from the scene

- complexity of the location

- weather and lighting conditions

- rail lines are bi-directional

- communication method and evacuation signals to be communicated to all Fire and Rescue Service staff

- the audibility of any message or signal

- the noise level at the scene

- the risk to the safety officer

- footprint of the incident.

8B.73   High speed trains and international services may take over a kilometre to stop.

Speed	Distance covered in 30 seconds
60 mph	805 metres
55 mph	738 metres
50 mph	670 metres
45 mph	603 metres
40 mph	536 metres
35 mph	469 metres
30 mph	402 metres
25 mph	335 metres
20 mph	268 metres

ACTIONS TO BE TAKEN BY SAFETY OFFICERS ON SEEING AN APPROACHING RAIL VEHICLE

8B.74 The rail industry has standard signals for stopping approaching rail vehicles.

8B.75 During daytime the signal is to raise both arms above the head.

8B.76 At night or in darkness the signal is to repeatedly and quickly wave a torch from side to side, at the train.

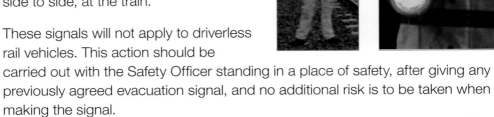

8B.77 These signals will not apply to driverless rail vehicles. This action should be carried out with the Safety Officer standing in a place of safety, after giving any previously agreed evacuation signal, and no additional risk is to be taken when making the signal.

8B.78 The rail system may mobilise a manager to perform the role of Responsible Person at Silver. This individual will assist with the identification of rail specific hazards, and may provide options for the removal or reduction of hazards.

8B.79 Safety officers should also consider the effect that unusually large volumes of water, for example from flooding or high volume pumping, may have on the rail infrastructure. Such volumes of water can

- weaken embankments
- misalign running rails.

8B.80 These consequences can have disastrous outcomes some time after the event. Safety officers and Incident Commanders should take action to prevent large volumes of water being directed onto rail infrastructure during incidents that involves unusually large volumes of water, if reasonably possible to do.

8B.81   Because of the potential for such damage to occur rail systems should not be used as a conduit for the removal of large volumes of water from nearby flooding incidents.

3.7 Actions of deployed crews

8B.82   Fire and Rescue Service personnel working on a rail line must be aware that in addition to any control measure implemented to protect the scene, a general awareness of the hazards likely to be involved must be maintained and appropriate control measures implemented.

8B.83   At the scene of an incident those undertaking operations at a bronze level will be responsible for reassuring the public and taking measures to protect people from further harm. This may be achieved by:

- providing reassurance, information and instruction to members of the public

- relieving members of the public or rail staff from any spontaneous rescue operations, when reasonable to do so

- keeping alert. 'stop, look, listen' before moving about the infrastructure

- be aware of safety signage or information provided for Fire and Rescue Service crews

- providing an initial assessment of the situation and regular updates to the Incident Commander

- informing the Incident Commander of any control measures implemented by rail vehicle drivers or rail staff

- use equipment and facilities designed for access and Fire and Rescue Service purposes

- making an assessment of surrounding hazards before using Fire and Rescue Service resources

- even after power off avoid unnecessary contact with rails and traction current equipment

- assess overhead line equipment and other cables for stability/damage

- remain aware of the situation and any changes you make to the scene, to assist with any future investigation

- following significant incidents, it may be useful for individuals to record observations and any actions taken, as soon as is reasonably possible.

3.8 Determine firefighting tactics

8B.84 When undertaking and directing firefighting operations at rail incidents consideration should be given to:

- use of relevant firefighting media and procedures appropriate to the hazards and risks present

- direction and proximity of crews, firefighting jets and equipment to live rails, or un-earthed cables

- the conditions crews are working in

- the damage caused to the infrastructure, for example arches, tunnels, overhead line equipment

- hose lines traversing rail, not covered by control measures

- maintaining an awareness of safe working area on rail infrastructure

- use of any tactical plans or built in facilities

- ascertaining the status of any automatic system to assist firefighting

- establishing and maintaining agreed communication procedures.

3.9 Carry out rescues

8B.85 In ideal circumstances, to prevent accidental contact of clothing or equipment, Fire and Rescue Service operations should be conducted three metres away from any live electrical traction current where possible. However it is recognised that in many rescue situations it will be necessary for the Fire and Rescue Service to work much closer to rail lines and to enter carriages.

RESCUES INVOLVING OVERHEAD LINE EQUIPMENT

8B.86 If a person to be rescued is within 1 metre of Overhead Line Equipment or they or any equipment to be used may come into contact with the overhead line equipment during the rescue Incident Commanders should implement control measures 3 and 4 (section 3.2 above).

8B.87 The only exception to the above is where the Fire and Rescue Service are called to deal with a rescue of a person or persons in contact with live overhead line equipment and there is an immediate life saving opportunity, and to wait for it to be earthed may lead to loss of life. In these circumstances Incident Commanders must request 'power off', and when confirmed crews wearing full personal protective equipment including dry gloves can remove any casualties from cables or machinery using dry non-conductive equipment.

8B.88 Additional factors that may require overhead line equipment traction current to be shut down include:

- if a rescue is to be carried out in smoke or high humidity (e.g. following a fire in a tunnel)

- there are signs of damage to or collapse of the overhead line equipment structures

- if it cannot be assessed how far a casualty is from traction current.

RESCUES INVOLVING DC ELECTRIC RAIL (THIRD AND FOURTH RAIL SYSTEMS)

8B.89 When performing a rescue from traction current involving electrified third or fourth rail systems Incident Commanders should implement control measures 3 and 4 (section 3.2 above). If necessary a rescue may be attempted before power off is confirmed providing:

- crews wearing full personal protective equipment including dry gloves; and

- the rescuer is standing on dry non-conducting material (for example dry clothing, wood, thick rubber).

8B.90 If this cannot be achieved then the person should be moved away using dry non-conductive material. Metal objects must not be used.

RESCUES FROM RAIL VEHICLES

8B.91 Incident Commanders should ensure that a systematic search of any rail vehicles involved and surrounding infrastructure is undertaken.

8B.92 When searching or rescue operations are underway at larger incidents, care should be taken when assigning identifiers to sectors or carriages involved. If more than one rail vehicle is involved Incident Commanders should consider making each vehicle a different sector. It would be useful to number each carriage to assist identification. It is important to ensure that any numbering system applies consistently when viewed from either side of the incident.

8B.93 Any identification system used should be shared with other agencies present.

8B.94 Extrication and evacuation of casualties can be resource intensive, requiring specialist Fire and Rescue Services resources, perhaps both locally and nationally. Operations can be protracted and challenging. In addition to crews knowledge information should be sought from the Responsible Person at Silver to identify the type of vehicle involved and any features or facilities provided for fire or rescue purposes.

8B.95 Access to vehicles can be difficult owing to the features of the infrastructure, the nature of the incident, the type of vehicle involved, and any vehicle security measures provided. It is not usual for rail vehicles to be fitted with areas for cutting away in order to rescue survivors. Research has been carried out into the best type of glazing to provide for passenger vehicles. It has been identified that glazing designed to allow passengers to break glass to escape also fails during impact, raising casualty numbers. Consequently the strategy in most modern rail vehicles is for the structure to provide adequate protection for passengers, resisting the effects of impact or derailment, while keeping the passenger contained in the vehicle. This should then enable personnel to escape or be removed through passenger doors.

8B.96 Fire and Rescue Service personnel engaged in rescue efforts should normally use the vehicle doors for access and egress.

8B.97 Rail systems may provide specialist teams that may support the Fire and Rescue Service's operations.

8B.98 The assistance that can be provided includes:

- specialist knowledge of rail vehicle

- specialist cutting and lifting equipment

- lighting facilities

- welfare and refreshment facilities.

8B.99 The evacuation of passengers, including those who may have restricted mobility, that do not require rescue, is the duty of the train operating company and infrastructure managers.

8B.100 Incident Commanders should be aware of the evacuation strategy for the rail system or location, and should enquire how any evacuation is progressing.

3.10 Resolve hazardous materials (Hazmat) issues

8B.101 Standard hazardous materials procedures will be broadly applicable to any hazardous materials incidents occurring on a rail system. Further information on resolving hazardous materials incidents is provided in the relevant national guidance and generic risk assessments.

8B.102 Hazardous materials incidents occurring on rail systems will predominantly be as a result of the nature of goods being carried or risks contained in the construction of the rail infrastructure or vehicle. Additionally, hazardous materials may be deliberately introduced onto the rail system.

8B.103 The nature of the incident will inform the Incident Commander's actions when determining response in terms of resources, deployment and crew safety.

8B.104 Information on the hazards associated with freight or vehicles can be obtained either through the infrastructure managers, for example the Total Operating Processing System or through the Responsible Person at Silver, if available at the incident.

8B.105 Incident Commanders should ordinarily consider the effect that the incident or firefighting operations may have on the community and environment. Such as:

- smoke or fume plumes

- the effects of any discharge from tunnel ventilation systems

- the effects of fire water runoff on near-by water courses, particularly in more remote/rural locations.

3.11 Establish effective liaison systems

8B.106 The management and resolution of rail incidents often requires the combined efforts of a number of agencies. It is therefore essential that effective and proportionate liaison is established at an early stage at the vast majority of rail incidents to ensure that the priorities and risks relating to different agencies are effectively managed.

8B.107 Interagency Silver meetings are particularly important at rail incidents. The early and regular exchange of information on the progress of emergency responders, assessment of the impact of an incident on the community and rail industries involved along with comprehensive reviews of safety issues can help to coordinate response and reduce the impact of the incident. In addition to normal protocols for interagency Silver meetings, the following points should be considered at rail incidents:

- At certain incidents the Rail Accident Investigation Branch or appropriate police service will look to commence investigation, in addition to any fire investigation required by the Fire and Rescue Service into the cause of the ignition and spread of a fire. As some evidence at rail incidents can deteriorate reasonable efforts to facilitate any statutory investigation should be made as early as possible

- The effect or implications of any near-by rail or transport incidents

- Identifying if any responders have designated the incident as a 'major incident'

- Identifying any communications issues, and assessing the impact

- The interagency Silver meeting should also be used to confirm with the Responsible Person at Silver the extent of the cordon in place, the current status of rail specific hazards and the availability and use of specialist equipment

- Appropriate reduction of the cordon to reduce the community or rail industry impact of the incident

- The progress of any rail industry managed evacuation and the impact on Fire and Rescue Service operations.

Phase 4 – Implementing the response

Phase 4 – Actions	• Communicate
Implementing the response	• Control
4.1	Implement effective control measures
4.2	Implement effective firefighting and rescue operations
4.3	Communicate the tactical plan
4.4	Implement deliberate reconnaissance to gather further incident information
4.5	Communicating with other agencies
4.6	Controlling the tactical plan

Considerations

4.1 Implement effective control measures

8B.108 Incident Commanders should ensure that appropriate and proportionate control measures are requested from the relevant infrastructure managers, using the most expedient method. Normally this will be requested and confirmed by radio transmission, via Fire and Rescue Service Control.

8B.109 It would normally be desirable to obtain confirmation from the infrastructure managers that the control measures have been implemented before committing personnel on or near the railway.

8B.110 In extreme circumstances, personnel may be committed before confirmation has been received. In these circumstances additional, temporary, control measures must be applied.

8B.111 In addition to safety officers being appointed for conventional Fire and Rescue Service purposes, at a railway incident a safety officer may also be appointed to watch for approaching train vehicles.

8B.112 For high speed or international rail services it is unlikely that any temporary control measure would provide sufficient protection from approaching rail vehicles.

8B.113 Fire and Rescue Service crews should receive an appropriate briefing on the actions to be taken and the firefighting, rescue or environmental protection tactics to be employed. The briefing may also include:

- significant hazards and risks

- actions to be taken on the sounding of any warning or evacuation signal

- safe working areas where operations are to take place

- any fixed installations available and how they provide protection or are to be used

- the role of other agencies working with Fire and Rescue Service personnel.

8B.114   When working alongside other agencies within any inner cordon information on significant hazards and control measures implemented should be shared with other responders. The emergency/evacuation signal and the appropriate response should be agreed with all parties working on or near the railway.

4.2 Implement effective firefighting and rescue operations

8B.115   Many rail incidents will be challenging and dynamic and outside the operations familiar to most Fire and Rescue Service personnel. Therefore particular care and consideration by the Incident Commander will be required when implementing the tactical plan, possible considerations will be:

- correct resources and equipment appropriate to the tasks

- an appropriate briefing to Fire and Rescue Service personnel, paying particular attention to the hazards identified with this incident

- the limited experience of Fire and Rescue Service crews with working on rail vehicles and infrastructure and relevant tactics.

- tactical plan needs to be realistic and achievable, balancing risk against benefit

- plan may include joint working with other agencies e.g. HART/specialised police departments

- the plan needs to comply with Fire and Rescue Service standard operating procedures and Fire and Rescue Service policies, based on *Generic Risk Assessment 4.2 Incidents Involving Transport Systems – Rail*.

4.3 Communicate the tactical plan

8B.116   The plan will need to be communicated to all Fire and Rescue Service personnel attending an incident, all other category 1 responders and attending agencies in line with Incident Command System protocols and local major incident procedures where appropriate. Communication should:

- confirm understanding of the tactical plan with Fire and Rescue Service personnel and broadcast tactical mode to all Fire and Rescue Service personnel

- disseminate information to other responding agencies and confirm understanding with reference to tactical plan and identified hazards of the incident

- implement liaison protocols/procedures with other emergency services to assist in the communication of Fire and Rescue Service activities and other agency activities

- consider skills, knowledge, competency, capabilities and resources of other agencies.

4.4 Implement reconnaissance to gather further incident information

8B.117 Reconnaissance should be carried out by suitable Fire and Rescue Service personnel to assist the Incident Commanders in evaluating the extent and nature of the incident, Utilising rail industry personnel to gain additional information may be of paramount importance to develop safe systems of work and to ensure safety for all category 1 and 2 responders working within the inner cordon, when formulating and implementing the tactical plan. Further information may be available from:

- premises information boxes
- infrastructure managers
- station control rooms
- responsible person at silver
- train operator's representative
- freight operator's representative
- Police Service or British Transport Police responders, including specialist response teams
- ambulance personnel, including specialist response teams
- total operating processing system information for national rail infrastructure vehicles
- detection, identification and monitoring/hazardous materials advisor
- urban search and rescue teams
- inter-agency liaison officer
- railway emergency teams
- Rail Accident Investigation Branch
- Her Majesty's Inspector of Railways from the Office of the Rail Regulation
- rail industry engineers
- rail vehicle manufacturers
- passengers.

4.5 Communicating with other agencies

8B.118 Incidents may require a multi-agency response to achieve a satisfactory conclusion. Historically communications both internal and external have been identified as areas of weakness in post incident investigations and debriefs. Therefore, Fire and Rescue Service Incident Commanders should consider carefully their methodology for communication with other agencies (category 1 or 2 responders). Areas for consideration will be:

- AIRWAVE radio communication system utilising interagency radio channels
- danger of reliance on mobile telephone networks
- the use of field telephones between emergency service control vehicles
- the use of runners if appropriate
- the use of interagency liaison officers
- the use of any mutually agreed method to overcome local difficulty, e.g. fixed communication systems on stations
- identification of adequate area for co-location of command vehicles
- the use of interagency Silver meetings to confirm incident situation and communication inter service communications structures
- the availability of satellite communications systems.

4.6 Controlling the tactical plan

8B.119 Once the tactical plan is in place, the Incident Commander must ensure that effective arrangements are in place to monitor the implementation, application and progress of the plan to ensure the objectives are being met. This will include the establishment and maintenance of:

- an appropriate command structure
- effective safety management systems
- appropriate communications systems
- effective arrangements for liaison.

Phase 5 – Evaluating the response

Phase 5 – Actions Evaluating the response	• Evaluate the outcome
5.1	Obtain and utilise specialist advice
5.2	Assessment of safe system of work
5.3	Evaluate the effectiveness of the tactical plan
5.4	Consider the appropriate and timely reduction of the size and impact of cordons

Considerations

5.1 Obtain and utilise specialist advice

8B.120   Incident commanders should not underestimate the range of expertise that is available from those detailed in section 4.9 above, and should use all sources available to assist with evaluating progress. This should include using the industry experts to confirm that the decisions and assumptions within the tactical plan remain valid and appropriate for the incident type, size, scale, location and associated hazards.

8B.121   Any new information or change in circumstances will require the Incident Commander to evaluate the impact of that information on identified objectives and as such the tactical plan may need to be amended.

5.2 Assessment of safe system of work

8B.122   Safe systems of work that are implemented as control measures to protect Fire and Rescue Service personnel and possibly other responding agencies working in the same location, should be continually re-assessed with consideration of the following:

- the information received from those listed in paragraph 4.9 above confirms the justification of continued Fire and Rescue Service activity

- the likely loss or impact caused by an incident

- the potential for escalation of the incident, likely involvement or spread of fire, hazardous materials

- a change in weather or any flood water conditions may have an affect on the suitability of the safe system of work

- the positioning of the rail vehicles and their stability as the incident develops, being affected by fire service operations or the result of fire/crash damage

- the functionality of railway fixed installations and support equipment, including Fire and Rescue Service communication facilities.

5.3 Evaluate the effectiveness of the tactical plan

8B.123   With all tactical plans there will need to be a continuous review of the priorities and objectives of the plan balanced against the risks being taken by Fire and Rescue Service personnel. Undertaking this review the following questions may be considered:

- Is crew safety and welfare being maintained?

- Are the risks being taken by Fire and Rescue Service personnel proportional to the benefit?

- Have comprehensive analytical risk assessments been completed and appropriate control measures implemented?

- Are the resources appropriate and adequate to achieve the tactical plan?

- Has there been a change in the level of risks pertaining to the incident, for example rescues carried out?

- Has any review of the plan required a change to the level or extent of control measures

- Have the operational tasks achieved the tactical plan, if not why not and what needs to be altered to achieve the tactical plan?

5.4 Consider the appropriate and timely reduction of the size and impact of cordons

8B.124 As part of continual evaluation, Incident Commanders should regularly review the size and appropriateness of any cordons with a view reducing the size and impact to the minimum needed to ensure safe and discreet operations. Some considerations may include:

- Obtain information on the progress of Fire and Rescue Service actions and any evacuation of members of the public by the infrastructure managers.

- Establish early Silver meetings to:

  – develop effective joint plans to mitigate the impact of the incident; and

  – agree handing over process to the appropriate organisation, agency, or company

- develop any media strategy and provide information to the public.

8B.125 Gathering information from responders and specialist advisors can assist an Incident Commander to evaluate the response. For a railway incident this may include:

- safety officers

- hazardous material and environmental protection officers

- interagency liaison officer

- press liaison officer

- infrastructure manager

- other emergency service responders

- on-site

- health and safety professional or statutory investigators (Health and Safety Executive, Office of Rail Regulation, Rail Accident Investigation Branch)

- rail industry engineers or rolling stock manufacturers.

Phase 6 – Closing the incident

Phase 6 – Actions	
Closing the Incident	
6.1	Scaling down Fire Service operations
6.2	Handover/ownership of railway
6.3	Facilitate debriefs
6.4	Facilitate post incident reporting
6.5	Maximise learning

Considerations

6.1 Scaling down Fire and Rescue Service operations

8B.126 This is an important phase of the incident and statistically a phase when accidents and injuries are prevalent. There is therefore a need to maintain effective command and control throughout this phase of the operations, which is likely to include:

- continued dynamic management of risk and a record of Incident Command decisions

- scene preservation in conjunction with advice from police/Rail Accident Investigation Board

- decontamination of equipment and personnel

- personnel welfare

- safe recovery of Fire and Rescue Service equipment.

6.2 Ownership/handover of railway

8B.127 At the end of a Fire and Rescue Service operation it will be necessary to identify on-going ownership of the incident scene, i.e. Highway Agency, local authority, private enterprise, railway authority, in order that correct handover procedures can be put in place.

8B.128 In most cases the railway will be handed back to the infrastructure manager, in this instance the Fire and Rescue Service Incident Commander should inform the infrastructure manager on scene and via Fire and Rescue Service Control that Fire and Rescue Service operations are complete, this will indicate that all Fire and Rescue Service personnel and equipment are now outside of the hazard zone (i.e. 3 metres). The railway is now handed back to infrastructure manager.

8B.129   Where a handover of command of the incident takes place from the Fire and Rescue Service Incident Commander to a responsible person from another agency, such as the police, or Rail Accident Investigation Branch there must be a full and thorough exchange of information which should be recorded at a Silver meeting as part of the ongoing risk assessment process (see appendix 1), and should include the following:

- the current Incident Commander

- the identification of the responsible person taking over the incident.

- who has track possession

- geographical extent of the possession

- the risk assessments in place

- safe systems of work being employed

- actions that have been taken including rescues and number of casualties, firefighting etc.

- what actions are currently taking place

- any personnel still deployed, and what agencies they are from

- any equipment still deployed

- location of any hazardous materials

- hazardous or unsafe structures

- environmental considerations

- contact details of relevant agencies that may be required to bring the incident to a satisfactory conclusion.

6.3 Facilitate debriefs

8B.130   As with similar types of major incident the Incident Commander will need to ensure the relevant records and information are made available for internal, inter-service and inter-agency post incident debriefs, which may include:

- on scene hot debriefs

- structured Fire and Rescue Service internal debriefs

- structured multi-agency debriefs

- critical incident debriefs (trauma aftercare).

6.4 Facilitate post incident reporting

8B.131   All railway incidents will be subject to some degree of post incident reporting. The extent and detail of any reporting will depend on the scale and severity of the incident. Compilation and circulation of multi-agency major incident reports may be determined by the strategic co-ordination group in line with National Policing Improvement Agency guidance for emergency procedures.

8B.132   Internal reports or other documentation may be disclosable and may be used in coroners or criminal court proceedings. Incident Commanders and the Fire and Rescue Service should consider the need for the following to be created and maintained during any incident and make appropriate arrangements for the security and availability of this information following any incident. Information sources may include:

- contemporaneous notes and/or statements from Fire and Rescue Service personnel
- continuous record of Fire and Rescue Service mobilisations and messages
- decision logs
- internal Fire and Rescue Service investigations and reports
- incident recording systems.

6.5 Maximise learning

8B.133   Fortunately serious rail incidents are rare and therefore when these occur Fire and Rescue Services should seek to maximise the benefits to the Fire and Rescue Service as a whole. The Fire and Rescue Service should consider the following as opportunities to measure and benchmark performance, identify potential for improvements and share lessons learned.

- interagency liaison officer operational guidance
- Fire and Rescue Service intervention strategies
- Fire and Rescue Service policies and standard operating procedures
- Fire and Rescue Service training
- equipment failures and successes
- lessons learnt and shared with other authorities and the Fire and Rescue Service.

Part C
Technical considerations

National Rail systems

General

8C1.1   The rail system within the United Kingdom is a complex infrastructure, comprising of different companies managing separate, but normally integrated, rail infrastructure.

8C1.2   The infrastructure managers maintain a safe rail infrastructure, to enable the transport of passengers by train operating companies, or freight by freight operating companies.

8C1.3   In some areas the company operating the train vehicles and the infrastructure manager may be the same organisation.

8C1.4   Rail transport can be divided into the following groups:

- Commuter/passenger metro systems, for example London Underground or Tyne and Wear Metro

- Commuter/ passenger main line, these form the majority of rail services within the United Kingdom

- Freight services, transporting goods

- High speed international service, connected to the European mainland

- High speed domestic service, utilising the international service infrastructure to transport commuters/passengers on high speed trains

- Industrial railways, used to transport goods, materials or workers. These may be found, for example, at large construction-sites, factories, dockyards. These may be connected to the wider national rail system, for the transport of goods and materials

- Heritage railways, historic rail systems maintained as businesses or by hobbyists for tourism or interest. These may be connected to the wider national rail system.

Network Rail

8C1.5   Network Rail is the infrastructure manager for the national rail system in England, Scotland and Wales. It owns and operates the railway track, power supply, signalling systems and owns many of the stations and depots.

8C1.6   Network Rail is responsible for providing and maintaining a safe rail infrastructure in order to allow other companies (i.e. Train Operating Companies) to safely transport passengers and/or freight. Train operating companies may also lease stations and depots from Network Rail and they will generally provide their own staff at these locations.

8C1.7   As infrastructure managers, Network Rail is responsible for safe operations within the infrastructure under its control. They may also be contracted to operate the rail infrastructure on other's rail systems, including some metro systems on behalf of a strategic body.

Freight operations

8C1.8   The national rail network carries freight including goods and materials being imported and exported. The vehicles carrying freight are owned and operated by freight operating companies. These companies will have emergency procedures for informing the infrastructure manager of an incident. Freight being transported may include:

- non-hazardous goods, products or materials
- the full range of hazardous goods including:
  - military explosives and munitions
  - radioactive materials.

8C1.9   Incident Commanders can obtain information on hazards associated with national rail freight wagons by requesting total operations processing system information via Network Rail. By providing the carriage number of the rail vehicle involved a check will be made against information, provided by the freight operating company on the type and quantities of goods carried.

Passenger services

8C1.10   Train operating companies will have their own emergency procedures for the scene safety management or evacuation of passengers from passenger trains. Once passengers have evacuated from a train they then come under the control of the infrastructure manager, who will cooperate with the train operating company to ensure passenger safety. In most emergency circumstances the train operating company will endeavour to keep passengers safe on the rail vehicle, looking to transport them to a place for safe evacuation (station, evacuation point, transferring to another rail vehicle).

8C1.11   In the event of an evacuation many train operating companies will provide facilities, including reception centres, for handling and assisting evacuated passengers.

High speed and international service

8C1.12 International train services operate in the United Kingdom on the 109 kilometres (68 miles) rail system linking the Channel Tunnel at Folkestone in Kent with St Pancras International Station in London.

8C1.13 The system is controlled from a rail control centre in Ashford (Kent). The High Speed Railway (HS1) rail system runs through the counties of Kent and Essex and then, onto London.

8C1.14 A domestic service operates on the same system, using high speed vehicles, similar to the international service.

8C1.15 It is intended that future development of the high speed service will extend this international rail system to other parts of the United Kingdom.

8C1.16 The international service rail infrastructure consists of purpose built 'long tunnels', tunnels, viaducts and bridges. During the planning and development of the system facilities where provided to support the operational intervention strategy for Fire and Rescue Service crews.

8C1.17 Emergency plans are maintained that reflect the intervention strategy developed during the planning and construction stage in partnership with all emergency response organisations, train operating companies and local and unitary authorities. Details of the tunnel construction and intervention tactics can be found in the Operational Guidance for Tunnels and Underground Structures.

8C1.18 The High Speed Railway (HS1), although built to the United Kingdom "standard gauge", has some unique features which significantly differ from the remainder of the United Kingdom's railway system.

8C1.19 These include:

- high speed railway – 300 kilometres per hour (186 miles per hour)

- no line-side signalling – Automatic bi-directional signalling system with data transfer via running rails into the train

- "People-less Railway"– With no access into the hazard area or line crossing when the railway is operating

- a generic plans book has been produced for the High Speed Railway (HS1) railway so that a common reference document is used by all users and emergency response organisation's who may be required to respond to the railway.

PART C–2
Metro systems

General

8C2.1   Metro systems principally operate to provide an exclusively passenger service, normally in a built up area. There are a number of towns and cities throughout the country that are served by metro systems. These can be owned and operated by the same company including ownership of the rail vehicles or managed by an infrastructure manager on behalf of a strategic body.

8C2.2   Metro systems exist in many of the UK's major cities including Liverpool, Newcastle and London and are a key component of local transport infrastructures. London Underground, for example, carries more passengers daily than the entire national rail system, and is key to the functioning of the capital and some surrounding areas.

8C2.3   Metro systems assist economic development and help reduce traffic congestion and pollution, forming an important part of a regional strategic transport policy. Metros can serve other transport systems including international ports and regional transport hubs. Many areas have invested heavily in new or the modernisation/expansion of existing systems.

8C2.4   Some metro systems may consist of driverless rail vehicles. In these instances the vehicle is automatically controlled. Any rail staff provided on automated vehicles normally have some responsibility for initiating emergency procedures and the evacuation of passengers, including the mobility impaired.

8C2.5   Although there are regional differences, most metro systems will have a number of common features and incorporate similar operating procedures.

Light Railways (Docklands Light Railway)

8C2.6   The Docklands Light Railway is an example of an automated metro system whose operating infrastructure includes tunnels and raised sections. The stations tend to be raised above street level, and are only staffed at sub surface locations. This is for emergency management reasons. The system is owned by the strategic transport authority in London, but is operated by concession holding companies, with responsibility within a geographical area of the system. Each of these concession holders will develop separate but integrated emergency procedures.

8C2.7   As with most modern rail systems, the day to day operations are highly dependant on computerised systems, close circuit television and other modern technical installations.

8C2.8   A standard light railway train can carry up to 250 people and other key benefits include:

- rapid acceleration

- relatively quiet in operation

- quick and safe braking.

8C2.9   Close liaison with the control room operators and Fire Service Control will be required to ensure that any request by the Incident Commander for example 'power off' is undertaken effectively.

8C2.10   In the case of Docklands Light Railway, the control room operators (rail) play an important role as their trains are driverless. A passenger services assistant will travel on every vehicle whose role is to ensure passenger safety and welfare, and implement emergency procedures. The passenger service assistant is also able to communicate with the rail system's control room and will be able to take charge of the vehicle should the need arise.

8C2.11   Other technical information regarding Docklands Light Railway:

- the traction power supply for sections or the entire railway can be turned off from a single control room

- Docklands Light Railway's trains have a Max Speed of 50 miles per hour (80 kilometres per hour) but normally operate at 40 miles per hour (64 kilometres per hour)

- there are emergency stop buttons at each station, these stop the trains, but do not cut the power to the third rail

- Docklands Light Railway's trains operate underground so are built to a high level of fire resistance

- there are fire extinguishers on each train that are accessible to passengers

- the trains have emergency egress handles that are accessible to passengers and emergency access handles that can only be operated by Docklands Light Railway's staff

- Docklands Light Railway's stations are not staffed, with the exception of the underground stations.

London Underground Limited

8C2.12 London Underground is a metro system and consists of some of the oldest and newest rail infrastructure integrated into the same system. Serving the capital and surrounding areas this extensive system typically operates:

- 520 trains in daily service

- 4.1 million passenger journeys per day.

8C2.13 Incorporated into the system is some of the oldest rail infrastructure, integrated into modern developments, which in turn is also integrated into wider rail and other transport systems. This includes running London Underground's trains along parts of national rail infrastructure. This can cause confusion to emergency responders when it is not immediately apparent who the infrastructure manager for the incident is.

8C2.14 In order to manage the risks a rigid London Underground management system exists to reduce or remove hazards. In line with most infrastructure managers London Underground look to ensure:

- the flammability of materials used on the system is as low as reasonably practicable

- the infrastructure is built or refurbished in consultation with emergency responders, with a view to improve response

- post incident debriefs take place in order to ensure lessons are learnt to improve service delivery for London Underground and responders

- close liaison with emergency responders takes place

- regular exercises are undertaken at Silver and Gold levels.

8C2.15 Emergency management arrangements mirror the structures established by Fire and Rescue Service responders.

PART C-3
Sub-surface railway incidents

8C3.1   A proportion of the rail infrastructure in urban areas is sub-surface or runs within long tunnels. This may include rail lines, depots, stations, service areas and shafts. Emergency response to these locations can be challenging and problematic. This guidance should be read in conjunction with specific generic risk assessment and guidance for incidents in tunnels and underground.

8C3.2   The facilities likely to be provided for Fire and Rescue Services operations at a sub-surface incident will vary depending on the age and complexity of the structure. At some sub-surface locations, e.g. underground rail stations, a control room will be provided and staffed by rail personnel.

8C3.3   For older infrastructure, undergoing redevelopment or significant improvement works, any sub-surface intervention strategy should be developed in consultation with the infrastructure manager, looking to use any existing facilities or proposed improvements. They should discuss with those responsible for the location the possibilities of improving facilities for Fire and Rescue Service response in areas where the construction does not reflect Fire and Rescue Service intervention needs.

8C3.4   At locations where the age, importance or complexity of the infrastructure has presented response challenges partnership working has developed innovative solutions to enable meaningful Fire and Rescue Services operations.

8C3.5   For new sub-surface developments or improvements to existing infrastructure the individual Fire and Rescue Service will look to develop an intervention strategy. Effective strategies will include consideration of the following features:

- A design size incident. This will be the starting point for determining the facilities to be provided for effective Fire and Rescue Service intervention. It will reflect the likely 'worse case scenario' for the Fire and Rescue Service to manage

- Establish agreed systems for control of hazards. These will not be reliant on rail personnel being in attendance and will normally be communicated via Fire and Rescue Service Control. If a Responsible Person at Silver is in attendance and has secured the requested level of safety, a message should be transmitted to control including the name of the representative providing the assurance

- Office of Rail Regulation communication facilities. This may include leaky feeder communications system, installed and maintained by the infrastructure managers

- Access locations, which may be stations, tunnel portals, intervention points, emergency response locations, and service shafts

- Any ventilation. This will include automatic or manual smoke control systems, natural ventilation facilities, or a 'piston effect' from other rail vehicles, that may impact upon Fire and Rescue Service operations

- Integration with the rail systems own managed evacuation procedures, to ensure operational response does not negatively impact on managed safety systems

- Clear understanding of the roles and responsibilities of responders including each rail companies own response and geographical responsibility

- Information gathering facilities. This may include:

  - premises information boxes available at stations or intervention locations

  - Responsible Person at Silver

  - station staff

  - predetermined plans kept available on the Fire and Rescue Service systems.

- Fire fighting water supplies. This may include mains provided throughout the sub-surface infrastructure for fire fighting purposes.

8C3.6 The Fire and Rescue Service should consider as part of any response to an incident in sub-surface locations the effect that the tunnel environment may have on the travelling public and staff. The high ambient temperatures and limited facilities for cooling passengers mean a significant stoppage of a rail vehicle may have an impact on passenger health and welfare. The severity of the impact may depend on the following:

- any over crowding on rail vehicles, particularly during rush hours

- failure of any battery operated cooling system due to prolonged main power stoppage

- dehydration

- climatic and environmental conditions

- if stoppage occurs in a tunnel ambient temperature and lack of air movement could be a factor.

8C3.7   Rail systems have arrangements in place to ensure that passengers are moved from affected trains in good time to prevent harm to travellers. The evacuation of passengers will cause significant delay and impact across the wider transport systems. In situations where harm to passengers may occur the Fire and Rescue Service will be called to assist. The Fire and Rescue Service should make clear that its assistance will only be called upon when there is the real potential or actual harm to the public, and not as a means to assist with wholly rail management problems alone.

8C3.8   The majority of modern passenger rail vehicles will have air conditioning which may increase the time that people feel comfortable within the tunnel environment. On occasions where traction power has been turned off the air conditioning will normally operate for only a short period of time. In these circumstances the air conditioning may be turned over to ventilation. Any contaminates within the tunnel may be drawn into the passenger's vehicle. On crowded rail vehicles the temperature can rise quickly with no cooling or limited air movement. If the situation causes batteries providing lighting and ventilation to fail, the likelihood of passengers taking independent action increases.

8C3.9   For further information on incidents in tunnels and underground refer to relevant national operational guidance and generic risk assessments.

PART C-4
Tramways

8C4.1 Tramway systems exist in a number of towns and cities across the country, typically the majority of tramway infrastructure is relatively new construction. However, there are some examples of older infrastructure such as in Blackpool and at various museums and heritage locations.

8C4.2 Modern trams are classed as light rail vehicles. A tram normally consists of a single unit which can be over 40 metres long with un-laden weight of up to 50 tonnes, which can be made up of several articulated sections.

8C4.3 Tram vehicles may operate in a rail environment, where there is physical or obvious separation between the rail lines and pedestrians or traffic. The same system may also operate in an environment where it moves among road traffic and/or pedestrians, in these circumstances the driver will comply with road traffic regulations. Trams have their own form of traffic signs and signals but these must also comply with the *Road Traffic Act and Road Traffic Regulation Act* and associated regulations.

8C4.4 So although trams comply with road traffic accident *regulations* they don't necessarily comply with the same signs and signals *on the ground* as road traffic [Fire and Rescue Service should also note that even in a "rail" environment, pedestrians have a greater degree of access than on a full railway, e.g. they are able to cross tracks at stations].

8C4.5 Trams can operate on different types of track, they can run on standard 'rails on sleepers' configuration or on grooved or tramway rails that sit flush to the ground, these are required for street running.

8C4.6 The vehicles are typically bi-directional with a full width cab at each end. The drivers' cab will normally have a deep windscreen and long side windows giving all around vision. The driver is positioned on the centre line of the cab with a wrap around console of controls such as radio, public address, heating, ventilation and lighting.

8C4.7 Modern tram bodies can be constructed of steel, or aluminium. Floors often consist of composite wooden sheets mounted on stainless steel with an abrasion resistant rubber covering. Trams windows are typically laminated glass bonded into the vehicle structure. From an operational perspective the Fire and Rescue Service should approach these in the same way as laminated train windows.

8C4.8 The majority of modern tramway systems are powered by a 750 volt (direct current) supply. The power is taken from overhead line equipment, which under normal operating conditions is permanently live. Overhead line equipment is normally carried on stanchions and is suspended at approx five to six metres above the ground. Where it may be necessary to reduce these heights, e.g. under a bridge, road traffic "safe height" warning signs will be used. The contact wire is made of copper of varying diameter and can be double contact wire.

8C4.9 On the roof of the tram is a pantograph, this allows the electrical power to be taken from the overhead line equipment and transferred to the electrical drive motors. The pantograph has an emergency isolation button either in the cab or elsewhere which upon actuation lowers the pantograph enabling the electrical supply to be broken. As with other rail systems the relevant control room can also isolate the power remotely in an emergency.

8C4.10 Modern trams will be capable of speeds up to 50 miles per hour (80 kilometres per hour) but most operate on the 'driver's line of sight' in the same way as a road vehicle, and speeds will be adjusted according to the location and surrounding conditions i.e. city or town centre sections will operate at much slower speeds due to the risk of collision with other road vehicles or pedestrians.

8C4.11 Trams as most rail vehicles may build up some impetus when travelling and there may be little friction between the wheels and the rails thus allowing a vehicle to 'coast' even when the power supply has been cut off. The potential danger this poses is that trams may still move a considerable distance through an unpowered section.

8C4.12 For operational Fire and Rescue Service purposes the management and resolution of incidents involving trams will largely be dictated by the environment in which the incident occurs. Fire and Rescue Service procedures for dealing with tram incidents in either a 'rail' or 'road' environment will be broadly appropriate, however, Fire and Rescue Service personnel should be aware that some local variations may apply.

8C4.13 The Fire and Rescue Service should work closely with operating companies to ensure that safe systems of work are adopted at operational incidents. This will include adequate liaison, pre-planning, training and familiarisation. Points to note during familiarisation should include for example; vehicle jacking points and inbuilt safety features such as battery and pantograph isolation points and the location and operation of such systems.

8C4.14 Where the Fire and Rescue Service have tram systems within their area they should also seek to ensure crews are familiar with how to contact control rooms, remote power supply isolation, location identification systems and earthing procedures which may vary between rail systems.

Industrial and heritage railways

Industrial railways

8C5.1 There are a number of different types of industrial railway, which are normally privately owned and can include large construction, gas, steel or colliery works, power stations, refineries and docks. The locomotives can be; diesel, battery, gas or electric powered and run on standard or narrow gauge tracks.

8C5.2 The railways can range from single line tracks that are only a few hundred metres long through to more complex multiple track systems with points and signalling, they may be several kilometres in length and cross over public roads.

8C5.3 Most industrial railway locomotives are used to haul quantities of materials or goods. They are primarily designed for power rather than speed and although some units are capable of greater speeds they will usually operate on-site at less than 20 miles per hour (32 kilometres per hour). Most industrial railways terminate at exchange sidings where main line locomotives will take over haulage of main line registered wagons. However, the Fire and Rescue Service should note that some local exceptions may still exist.

Heritage railways

8C5.4 The aim of these railways is to preserve as many historical aspects of the rail industry as possible, including infrastructure and steam or diesel powered engines. They are usually run by rail enthusiasts as a commercial concern and normally rely heavily on volunteers. Some railways operate a daily service throughout the year while some are seasonal or operate at weekends only. Local contacts for Heritage railways are referred to as the 'Responsible Officer of the Day'. Fire and Rescue Services who may attend heritage railways should maintain relevant contact details.

8C5.5 Many other aspects of the railway are the same or similar to the network rail infrastructure they are still subject to rail and other health and safety legislation. Fire and Rescue Service personnel should be familiar with systems in their area, paying particular attention to power systems and emergency arrangements.

Narrow gauge

8C5.6 There are a number of narrow gauge railways throughout the country. These tend to be heritage or industrial in use. Narrow gauge may also be found on major construction-sites, particularly during tunnelling or mining operations. In these circumstances the rail system is normally temporary in nature. The gauge refers to the distance between the running rails being narrower than on national, metro or light rail systems.

8C5.7 The terminology used and the remaining features for example points, signals, level crossings and location markers remain the same. Narrow gauge will not be directly connected to the wider rail network.

8C5.8 Although narrow gauge railways are generally much slower and smaller than for example network rail, there is still the potential for an incident that would require a response from the Fire and Rescue Service.

8C5.9 Fire and Rescue Service operations such as jacking and blocking can be more problematic at incidents involving narrow gauge railways because of the restricted access caused by the lower profile of narrow gauge rolling stock.

Other

8C5.10 Many other rail systems exist for particular purposes such as:

- local monorail systems

- people mover systems at airports

- funicular railway systems often used in tourist areas.

8C5.11 Fire and Rescue Service personnel should be familiar with any systems in their area, paying particular attention to power systems and emergency arrangements.

Specific operational considerations

8C5.12 Additional hazards that are associated with different types of heritage rolling stock can include:

- Liquefied petroleum gas cylinders [older buffet cars], where these exist they will typically be stored underneath the carriage. If involved in fire operational personnel should express caution when approaching, possibly using effective cover to direct cooling jets

- Steam vehicles will have a boiler with hot coals a high pressure steam boiler and a large fire fuel loading. These vehicles require specialist supervision whilst in operation, and additional advice should be sought on boiler safety

- Many heritage carriages are of timber construction often framed with steel exterior panels. Wooden superstructures are often painted with highly flammable varnish. Carriage roofs on older vehicles can be covered with layers of canvas and a linseed oil/chalk mix

- Heritage rolling stock is often furnished with traditional furnishings including in some cases extensive wood panelling, which will make a significant contribution to the fire loading

- Asbestos maybe present in heritage coaches which should be marked. If involved in fire or broken up this could present a severe respiratory hazard

- Within this category can also be found a number of rail systems that may be powered by alternative power sources, for example cliff water powered rail systems. For these systems the operational guidance contained within this note is broadly applicable, but local Fire and Rescue Services will have to examine and test this guidance within the specific context.

8C5.13 Other risks and hazards remain predominantly the same as any rail incident and it is the responsibility of each Fire and Rescue Service to pre-plan, assess the degree of risk and ensure that there is a safe system of work implemented It is particularly important that operational personnel have a good understanding of the different types of risks that these types of railways can present.

8C6.1 In the image below, two trains weighing around 1000 tonnes, are passing at a combined speed of 150 miles per hour. Fairly minor wind conditions can create enough noise to mask the sound of their approach and, because of the track curvature; they will only become visible when they are 300 metres away. This distance can be covered in eight seconds, so to cross the line now would be extremely hazardous.

8C6.2 Firefighters standing by the lineside cabinet, are less than one metre from the rail vehicle. Clearances are even tighter inside the tunnel. There is a short length of rail in the undergrowth and other trip hazards. A loose drain cover is also obscured. Any contact with the overhead power cable will result in injury or death.

8C6.3 Personnel expected to carry out operations in the railway infrastructure should have some knowledge of the general features they are likely to encounter. The information provided below is for guidance purposes only, and is based on the national rail network. It is the responsibility of the local Fire and Rescue Service to ensure that the information provided is appropriate and current to their local rail system.

Track layout

8C6.4 A running line is a line used by trains to go from place to place. Each running line will have a name, for example *'Up Main'* or *'Down Goods'*, as well as a speed limit. Picture 1 overleaf shows a railway with two tracks – one for each direction. Trains travel away from the camera on the left hand track (in this case the *'Down Main Line'*) and towards the camera on the right hand track (in this case the *'Up Main'*). The speed limit for this particular line is 75 miles per hour.

Picture 1

75 Down Main

75 Up Main

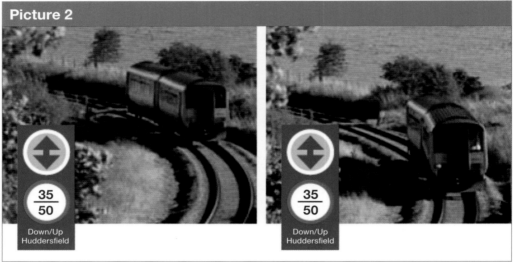

Picture 2

35 / 50 Down/Up Huddersfield

35 / 50 Down/Up Huddersfield

8C6.5 Sometimes trains can travel in both directions on the same track, as shown in Picture 2 (the *'Down Up Huddersfield Line'*). Here the speed limit depends on the type of train – either 35 or 50 miles per hour. Always assume the train will approach at the higher speed.

8C6.6 Some busy railways have four or more lines. In Picture 3 opposite, the two tracks on the left (the *Up and Down relief lines*) have a speed limit of 100 miles per hour. To the right (the *Up and Down main lines)* trains are allowed to travel at 125 miles per hour.

8C6.7 Rail vehicles generally run on the left, however, **rail lines can be reversible and rail vehicles can approach from either direction.** This can be because a line has, for example, stopped rail vehicles for engineering works somewhere along its length. Points and crossovers are used to allow the train to move between lines. Therefore the name 'up' or 'down' cannot always be predicted or relied on.

Picture 3

Up Relief   100 | Down Relief   100 | Up Main   125 | Down Main   125

8C6.8 Near busy stations the track layout can be complex, with many lines, points and crossovers.

Terminology

8C6.9 Terminology used in this document will be applicable to the vast majority of rail systems in the country. However terminology on rail systems has built up over many years and there can be significant local variations. The Fire and Rescue Service should use local familiarisation visits to develop understanding of any risk critical local variations.

8C6.10 The picture below shows the terms used to describe parts of the track.

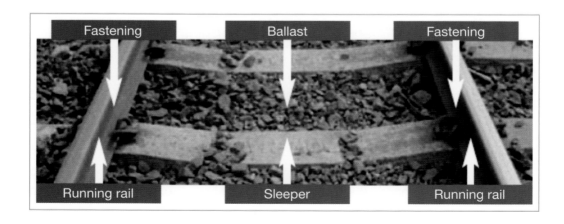

Fastening | Ballast | Fastening

Running rail | Sleeper | Running rail

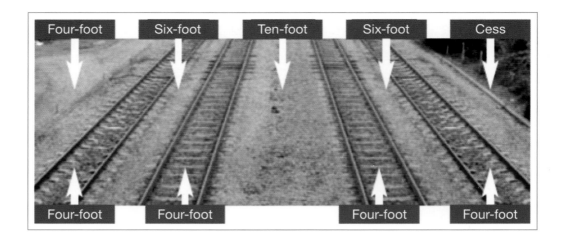

- running rails are the two rails that a rail vehicles wheels run on
- sleepers support the rails and keep them the correct distance apart
- fastening hold the running rails to the sleepers
- ballast keeps the track in place.

8C6.11 This picture above shows the terms used to describe parts of the railway.

- the cess is the area alongside the railway
- the four-foot is the space between the running rails of one line
- the six-foot is the space between a pair of lines if they're the normal distance apart.

8C6.12 Sometimes, if there are three of more lines, a wider space is provided between the two lines. This is known as the ten-foot or 'wideway'.

8C6.13 All these expressions are terms and must not be used as measurements.

Places of safety and refuges

8C6.14 There are terms that the rail industry uses to describe people's position in relation to the location of the track, for example 'lineside' or 'on or near the track'.

8C6.15 The nature of emergency responder work means that absolute rules are problematic in applying to all foreseeable circumstances. It is important that crews consider:

- the environment they are working in (for example, dark tunnels or flooded locations)
- any equipment that is being carried or used
- the type of infrastructure on which they are working (for example, high speed and international services require additional consideration and local planning)

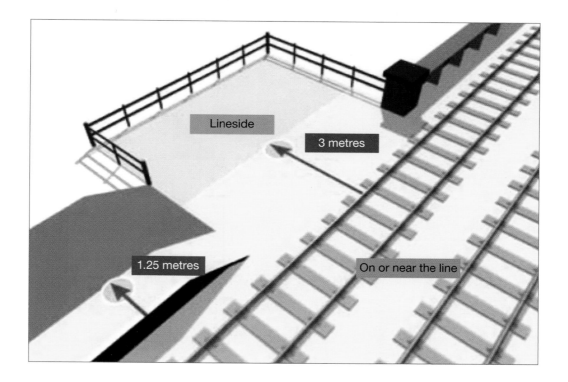

- visibility of crews to approaching rail vehicles

- the urgency of the task they are carrying out.

8C6.16 Some parts of rail infrastructures have designated 'authorised walking routes', providing safe access to or from a place of work. These are normally found near depots, siding or stations. Once control measures are implemented use of these facilities will assist operations. At larger incidents the designation of agreed temporary walking routes would assist safety management at an incident. This will normally highlight avoidable hazards, such as walking on line-side cable trough, not intended to be used as pathways.

8C6.17 On some parts of the railway, the space between the track and the nearest wall or structure is very narrow. These are areas of limited clearance. The signs shown below are examples. They indicate that there is no position of safety on this side of the railway for the length of structure beyond it.

8C6.18 A refuge is an area where it is safe to stand when a train is passing. Examples of refuge are shown above.

8C6.19 The examples of signs shown below indicate there are no positions of safety or refuges on this side of the railway, but there are on the other side.

8C6.20 This example of signage clearly indicates that the area beyond is too dangerous for rail employees whilst trains are running.

8C6.21 Any signage providing information on safety should form part of any operational risk assessment. This will include:

- careful consideration of the circumstances of the incident, particularly life risk

- appropriate and proportional control measures implemented

- and the resulting benefits of any Fire and Rescue Service action

- access and egress.

8C6.22 All these elements should be assessed, before personnel are committed.

8C6.23 When moving about the rail infrastructure the following points must be observed:

- Keep to the defined working area designated by the Incident Commander (in rail industry terms this is known as an 'emergency possession')

- Stay alert, keep watching and listening

- Do not assume you are safe because a signal is showing a red light or stop

- Use a designated walking route or pathway. Otherwise walk in the cess or, if necessary, the four-foot

- If possible face on-coming trains (remember tracks can be bi-directional)

- Stay within range of your safety officer and know what to do if you hear the agreed evacuation signal.

Safe working distances for Fire and Rescue Service operations

8C6.24 In circumstances where Fire and Rescue Service operations can be safely concluded whilst maintaining a distance of 3 metres from any live rail or overhead line equipment it is normally appropriate to apply control measures 1 or 2. (see Part B section 3.2) This will include many trackside fires, where incident commanders should continually assess the throw of firefighting jets to minimise risks to personnel.

8C6.25 Where operations need to take place closer than 3 metres to the track but no closer than 1 metre to any live rail, control measures 3 and 4 should be applied. In these circumstances it is possible that Fire and Rescue Service operations and or equipment could extend to closer than 1 metre to any live overhead line equipment. Incident commanders should therefore brief safety officers to monitor operations and if in doubt should request that the overhead line equipment traction current be discharged via earthing by rail professionals.

8C6.26 Where operations need to take place closer than 1 metre to any live rail, control measures 3 and 4 should be applied. In these circumstances Incident Commanders should also request that the overhead line equipment traction current be discharged via earthing by rail professionals.

8C6.27 The only exception to the above is where the Fire and Rescue Service are called to deal with a rescue of a person in contact with live overhead line equipment. In these circumstances, where there is a life saving opportunity, and waiting for the overhead line equipment to be earthed may lead to loss of life, Incident Commanders must request 'power off'. When confirmed crews wearing full personal protective equipment including dry gloves can remove any casualties from overhead line equipment using non-conductive equipment.

8C6.28 For reference from generic – standard operating procedures:

1. Inform the infrastructure manager of an incident on or near the railway.

2. Request rail vehicles are 'run at caution'.

3. Request that rail vehicles are stopped.

4. Request power off.

Action on an approaching a rail vehicle

8C6.29 If you see a rail vehicle approaching, for example if you are working in an area protected by trains 'running at caution, you should take the following action:

- Go to an agreed position of safety quickly and safely

- The driver of the rail vehicle will sound a rail horn as warning. Raise one hand above your head to indicate you have heard the warning

- If there are repeated blasts of the horn be extra careful as the rail vehicle may move in another direction

- Never assume you know which line the rail vehicle is on as it may be crossing points and be moving closer to you

- Keep watching the vehicle until it has passed

- Before leaving the place of safety ensure no other vehicles are moving nearby.

8C6.30 If, for whatever reason, you cannot reach a place of safety lie down and gather loose clothing under you. Personnel should not under any circumstances, lie down in the four foot.

Stopping a rail vehicle

8C6.31 Stopping of rail vehicles is normally via message to Fire and Rescue Service Control however in unusual circumstances an individual can stop a train by giving a danger signal. This must be clearly visible to the driver.

8C6.32 In daylight raise both arms above your head.

8C6.33 In darkness or poor visibility wave a light vigorously at the driver.

Moving about the rail infrastructure

8C6.34 If necessity requires you to cross the line directly ensure no trains are approaching and move straight across without stepping on rails or sleepers.

8C6.35 Take great care near points. Even after traction power has been isolated points can still move trapping feet (A).

8C6.36 On lines with conductor rails find a gap in the conductor rail (B).

8C6.37 Otherwise step over both the running rail and conductor rail together – never put your feet between them (C).

8C6.38 If provided and reasonably practicable cross at a place where protective guarding has been provided (D).

Using road vehicles near the line

8C6.39 Road vehicles can be a serious danger to rail vehicles if they are positioned near the line without proper care. The following points should be observed:

- no vehicle should be within 2 metres of where a rail vehicle may approach

- hazard warning lights should be on and in darkness headlights dipped

- only turn the vehicle at an appropriate turning point and keep the back of the vehicle furthest from the line

- once appropriate and proportional control measures are in place all red lights must be turned off.

PART C–7
Power systems

Electric

8C7.1   Electrified railways can operate under a number of systems and voltages, using the following traction power supply systems:

- overhead line equipment
- third rail supply
- fourth rail supply.

8C7.2   Some rail systems obtain their traction power by 'picking up' electricity from the overhead line using a roof mounted pantograph and some from third or fourth rail systems using collector shoes mounted close to the bogies.

Overhead line equipment

8C7.3   Overhead line equipment power can be provided for alternating current or direct current rail vehicles. Consisting of a live contact bar or wire suspended by a catenary wire, this is supported by a complex system of suspension cables, arms and tension devices. The parts that are normally live are highlighted in red below.

8C7.4   Every overhead line equipment structure has a unique number displayed that can be used for identification purposes. This can be extremely useful to the Fire and Rescue Service and infrastructure managers when determining the location of an incident.

8C7.5   The equipment is fed from a railway feeder or sub feeder station. These operate at up to 25 kiloVolts alternating current for "heavy" vehicles (national rail network, large metro systems). Typically "light" rail vehicles (trams, lighter metro vehicles) operate between 550-750 volts direct current but exceptionally may be up to 1500 volts direct current.

8C7.6   The rail vehicle will have a 'pantograph' on top to collect traction power. This will make contact with the live bar or wire, and will therefore also be 'live'.

8C7.7   Always assume that the overhead line equipment and everything in contact with it is live and extremely dangerous until formal assurance is provided.

8C7.8   Each overhead line equipment structure has a cable connecting it to the running rail. This is known as a bond. Some bonds are coloured red and are dangerous if they become disconnected. They must not be touched and should be reported to the rail infrastructure manager or Responsible Person at Silver, to ensure control measures are adequate.

8C7.9   The height of the contact bar or wire is normally between 4.7 and 5.1 metres but at low bridges it may be as low as 4.14 metres. In addition, at level crossings it is increased to 5.6 metres.

8C7.10 Overhead line equipment is under tension and therefore if damaged could collapse and recoil with force, remaining electrically charged until safely isolated and earthed.

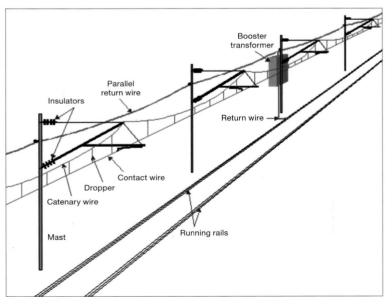

8C7.11 The overhead line equipment on national or metro services is divided into sections by means of switch gear at feeder stations and track sectioning cabins. It operates on sections or circuits that vary in length, but can be 30 miles or longer.

8C7.12 Incident Commanders should therefore be mindful that when isolation is requested it may have a serious impact on all train services within a radius of up to 15 miles. The effect of which could leave station platforms overcrowded and also trains left stationary on the track or in tunnels with no power supplies to keep lighting or air conditioning units functioning.

8C7.13 These circumstances are likely to raise the stress levels of stranded passengers, which could lead to a further risk of passengers descending from the train onto the track.

Third rail and fourth rail

8C7.14 The third rail traction system utilises a conductor rail operating at a nominal voltage of 750 volts direct current in most areas. A 'pick up shoe' on the train conducts the electrical current from the rail to the motor of the train. Some shoes are interconnected and can be live even if they are not touching the  conductor rail themselves. The return circuit is normally provided by the axles and wheels. The image below shows the live parts highlighted in red

8C7.15 On London Underground there is a fourth rail between the running rails which acts as a return circuit which can carry direct current voltage at 250 volts direct current. The live parts are shown below in red.

8C7.16 The control system is similar to that for the overhead line equipment. Sub-stations convert alternating current to the direct current supplied to the conductor rails and the overall contact of the circuits is supervised from a control room.

8C7.17 Qualified rail infrastructure staff can isolate local sections on-site by, for example, the manual operation of trackside switches. In either case the person providing assurance to the Fire and Rescue Service of isolation and providing details of the safe working area for crews should be recorded and passed via Fire and Rescue Service Control, for forward transmission to the infrastructure manager.

8C7.18 Anything touching the line, including fire fighting media, flood water or parts of the rail vehicle should always be regarded as live for both third and fourth rail systems, until assurance has been provided.

8C7.19 There is a risk that a rail vehicle may bridge an isolated and an energised section. This may inadvertently re-energise both sections. Deployment of short circuiting device by rail professionals, or confirmation of power-off should be sought.

Diesel

8C7.20   In the absence of third and fourth rails and overhead line equipment systems, rail vehicles will be self powered, this is almost inevitably by diesel traction.

8C7.21   Diesel vehicles may operate independently of any electrical power supply being available to the rail track or overhead catenary. When operating under diesel power rail vehicles can only be stopped by either communication with the driver or when the vehicle reaches a stop signal.

8C7.22   Diesel vehicles may carry significant amounts of fuel, lubricants and batteries, and locomotives and some carriages will be fitted with electric alternators, and electric traction equipment.

8C7.23   These vehicles can be used to move both passengers and goods.

Battery

8C7.24   Batteries are used as the sole traction power source on limited numbers of rail vehicles, including passenger carriages. This type of rail vehicle is typically used for maintenance work on London Underground and on some industrial and heritage railways. In future however hybrid battery-diesel vehicles could be encountered where traction power can be from either source.

Steam

8C7.25   Steam engines can be found operating on heritage rail systems and, to a lesser extent, as prestige passenger services on the national rail network. Incidents on these vehicles tend to be rare but may present additional hazards associated with:

- high fire load, including coal fuel and traditional passenger vehicles
- source of ignition
- steam and high pressure steam
- boiler and boiling water.

Dual voltage passenger freight

8C7.26   Rail vehicles that need to operate on systems with both overhead line equipment and third rail systems will be fitted with dual voltage facilities allowing pickup of traction current from either source.

Other rail system power supplies

8C7.27   Electrical supplies to signalling equipment and points are usually carried adjacent to the rail line in a concrete trough or gulley and will normally be supplied independently of traction current. This means that even after confirmation of power off has been received from the infrastructure manager, other electrical sources that are not traction current may remain live.

8C7.28   Fire and Rescue Service crews should be aware that damage to these cables can cause severe disruption to railway operations and present a major hazard to firefighters.

PART C–8
Utilities

General

8C8.1   Rail systems provide a convenient national conduit for utility company's delivery infrastructure including gas, electricity, communications and water.

Electricity

8C8.2   Cabling and infrastructure for national grid or local third party electrical supplies are likely to present the most immediate difficulties to firefighters when dealing with railway incidents. Local Fire and Rescue Services should identify the presence of additional electrical supplies when undertaking routine familiarisation visits

8C8.3   The infrastructure manager responsible for the system has no direct control over these supplies; however the Responsible Person at Silver may be able to provide information on the identification and ownership from their organisational database. The duty for isolating the supply rests with the utilities undertaker, and normal local Fire and Rescue Service procedures will apply.

8C8.4   Where high voltage power lines are located close to isolated overhead line equipment power cables it is important to ensure that the cables are isolated and earthed to avoid the introduction of additional risks to firefighters through the induction of electrical current.

8C8.5   Cable trunking is provided to carry cables through the infrastructure and may appear to provide a suitable pathway for access and egress for firefighters. Generally, most trunking is capped with a thin concrete slab. This has little tensile strength and can be susceptible to vandalism, often leaving it broken up and exposing cables. Trunking covers should not be used as walkways along the trackside as they are not designed for that purpose, and may fail. In darkness they can also present significant slip, trip and fall hazards.

PART C–9
Fixed structures on the railway

Railway sidings, depots, marshalling yards and industrial railways

8C9.1   For the purposes of this document these different types of rail premises can be considered the same.

8C9.2   They can be described as a short stretch of track or tracks that are connected to the main infrastructure. They may be provided for a number of purposes, for example:

- Parking of rolling stock

- Storage of rail vehicles, including those carrying hazardous materials

- Loading and unloading

- Maintenance facilities particular to the type of vehicle catered for such as fuel, chemicals, gas cylinders and waste

- Allowing trains to pass

- Sidings are often designed to allow several trains to be spaced closer than usual

- No signal may be provided

- Parking rail vehicles for servicing, cleaning or maintenance

- Loading or unloading goods

- The control and ownership may be shared between several parties. Extra care will be necessary to ensure the aspect of infrastructure requiring control measures to be applied are implemented.

8C9.3   These locations can present a challenge to the Fire and Rescue Service as various aspects and areas of their control may be under separate ownership and management control.

8C9.4   At some locations Fire and Rescue Services should identify the management and the location, uses and processes that occur at these locations, and develop appropriate detailed plans. These plans should identify areas of ownership and management responsibility such as:

- who operates in which area

- whose responsibility is it to provide safe access for the Fire and Rescue Service

- who controls traction current to which areas

- who, and to what extent are vehicles controlled

- who the Fire and Rescue Service contact to obtain information relating to hazards and risks, including what is stored on rail vehicles

- arrangements for attendance of Responsible Person at Silver

- means of identifying the Responsible Person at Silver and what authority will they have over the wider premises.

8C9.5 Railway wagons laden with hazardous materials, including explosives are occasionally parked in railway sidings or depots. Military explosives will not be transported with other goods.

8C9.6 Incident Commanders must be aware of the potential for the wide variety of hazardous loads that may be encountered in such locations and the importance of liaising with rail staff where available to identify loads.

8C9.7 Until a clear assurance has been provided that all train movements have been controlled Fire and Rescue Service personnel must not:

- attempt to go between two stationary trains, or a stationary train and a set of stop blocks, unless there is at least a gap of 30 metres between them

- crawl under or over any rolling stock.

Level crossings

8C9.8 Level crossings are railway lines crossed by a road or right of way without the use of a tunnel or bridge.

8C9.9 Types: Crossings are categorised into two main groups although the layout, configuration and use of level crossings vary from location to location, so each one is essentially unique:

- **Protected (Active) Crossings** – These crossings give warning of a train's approach to vehicle users and pedestrians through closure of gates or barriers, or by warning lights and/or sound.

- **Unprotected (Passive) Crossings** – These crossings have no warning system to indicate a train's approach. The road user or pedestrian is responsible for ensuring that they can cross safely.

8C9.10 Fire and Rescue Service involvement at these locations includes many instances of road vehicles being struck by rail vehicles. These incidents can require careful assessment in terms of applying appropriate control measures, and local crews should be thoroughly familiar with level crossings in their area. Without prejudice to crew or public safety the Incident Commander will need to consider the impact that a request has on the wider rail system.

8C9.11   If, for example, there is no life risk and no personnel, equipment or jets are going to come within three meters of the overhead line equipment the Incident Commander will not normally require the traction current to be isolated and earthed, only requiring that all rail vehicles are stopped. This will mean that Fire and Rescue Service operations can be conducted safely, and that the impact of the incident to the wider travelling community can be minimised.

Railway stations

8C9.12   A station usually consists of one or more buildings for passengers and/or possibly goods and may be constructed over a number of levels. A 'terminal' or 'terminus' is a station at the end of a railway line.

8C9.13   Large rail stations can have a number of rail companies operating in different areas. These may have different emergency procedures

8C9.14   Railway stations will have public and non-public areas. Areas for providing public access will generally present limited hazards. Some facilities provided to keep the public safe can present a potential obstruction to a Fire and Rescue Service, for example platform edge doors or barriers. Methods of opening these facilities should be known and readily available to the Fire and Rescue Service.

8C9.15   Non-public areas can present additional hazards to those generally encountered in public areas, for example:

- fast moving rail traffic

- high voltage electrical equipment for train stations and infrastructure

- traction current

- unusual direct access to the track.

8C9.16   However, in addition to the generic rail hazards encountered on other parts of the railway, firefighters should also be aware of other factors that may have a bearing on incidents located on or near a railway station.

8C9.17   Many railway premises are historic buildings and as such are likely to have heritage value. Some buildings including station houses and other ancillary buildings may have been granted listed building status.

8C9.18   Other buildings located adjacent to railway stations may include those for the purposes of warehousing or maintenance, consequently the potential for hazardous substances must always be considered when attending incidents on railway premises.

8C9.19 Larger railway stations will encompass a wide range of commercial outlets from newsagents and fast food outlets, to high street stores selling all manner of goods and services. Parts of the building may be given over to use by hotel, or offices, and may share access and exit routes, but remain otherwise separate.

8C9.20 Many railway stations have capacity for large numbers of people to be present, giving rise to issues such as:

- mass or partial evacuation of the premises
- widespread panic
- potential as a terrorist target
- mass decontamination.

8C9.21 During the planning and development or refurbishment of such locations the Fire and Rescue Service should promote exercises with local responders, including local authority emergency planners to examine the rail industry's and other stakeholder response to such events.

Bridges and viaducts

8C9.22 A 'bridge' is a structure built to span a gorge, valley, road, railway track, river or any other physical obstacle. The design of a bridge will vary depending on the function of the bridge and the nature of the terrain where the bridge is to be constructed. There are a number of additional hazards to consider when dealing with incidents in the vicinity of bridges. These can include working at height, restricted safety areas and difficult access.

8C9.23 Bridges that have been struck by moving road or rail traffic should be subjected to an examination by the infrastructures' management. All bridges and viaducts have unique identifying numbers that should be relayed to the infrastructure manager via Fire and Rescue Service Control, if no Responsible Person at Silver is present.

8C9.24 Viaducts are used to overcome steep gradients that are caused by geographical features such as gorges, valleys etc). They are raised sections of track, supported on pillars or on a series of arches. Viaducts present the same hazards as bridges but in addition the arches beneath (particularly in urban areas) are sometimes put over to commercial use, with each arch becoming a commercial unit.

8C9.25 Fires occurring in these locations can have a significant impact on the operation of the railway. These premises can sometimes contain commercial tenants who may store materials inappropriately. These locations, and secluded parts of the rail infrastructure can be frequented by drug users and vagrants with associated safety considerations for crews. Care should be taken when responding to such premises and Incident Commanders should consider the type of undertaking occupying the premises.

8C9.26 During the incident an infrastructure manager may wish to examine the integrity of the structure and any impact the incident may have had on the alignment of the rails. The Fire and Rescue Service should provide reasonable facilities for this activity at the earliest reasonable opportunity, in consultation with the Responsible Person at Silver. It will be necessary to identify the known hazards and risks and additional hazards brought about by Fire and Rescue Service operations, providing the infrastructure manager's representatives with the information.

Source: GMFRS

Tunnels

8C9.27 Rail tunnels can consist of a single bore with single track through to highly complex infrastructure containing multiple lines, with tunnels combining into multiple bi-directional routes, or incorporating underground sidings and depots.

8C9.28 An incident in a tunnel can present additional problems to operational personnel. This includes split attendances, length of hose lines, logistics of moving equipment, communications, breathing apparatus operations, difficult underfoot conditions and excessive heat/smoke.

8C9.29 Overhead line equipment may have also been 'brought down' and could be a hazard in any firefighting operations. In addition, limited clearance, poor lighting and difficult working conditions could increase the risks to personnel. The generic risk assessment and operational guidance for incidents in tunnels and underground structures should be referred to.

Points

8C9.30 A set of railway points is a mechanical installation enabling trains to be guided from one track to another. Points can be either mechanically or electrically moved from within a signal box or a control room.

8C9.31 Points can represent a hazard to firefighters as they can move without warning and therefore present a significant trap hazard. To avoid railway points becoming frozen and inoperable during cold weather, electrically powered point heaters are used. Personnel must always be cautious as the electrical supply to the points is independent of all other power supplies. If isolation is requested for a particular area of the track this may not include the power supply to the points.

8C9.32 Traditionally liquid petroleum gas was used to fuel point heaters. The gas cylinders were normally housed in small wooden cabinets next to the track, the main heating assembly would be supplied with gas via a length of low pressure tubing. Such examples may still be found on heritage or industrial railways.

Signalling equipment

8C9.33 Signalling devices are provided for the information of train drivers and rail staff only and should never be relied upon by Fire and Rescue Service personnel.

Location markers

8C9.34 Providing the exact location of an incident can be difficult due to a lack of reference points especially if the incident occurs in a rural location or information received from members of the public on the location of the incident may be inaccurate due to unfamiliar surroundings.

8C9.35 Assistance in identifying the location of an incident can be aided by the following:

- individual numbers marked on bridges and tunnels

- markings on signal gantries and signal boxes

- identifying letters and numbers on individual overhead line equipment support structures

- mile marker posts alongside all lines which are numbered consecutively.

Trackside telephones

8C9.36 Rail systems usually have a comprehensive telephone network enabling communications between control offices, stations, depots, most level crossings and signal boxes. These can provide additional communication routes, however it will be necessary to ensure adequate training is undertaken to ensure understanding of appropriate usage. Therefore Fire and Rescue Service operations should, whenever possible, communicate with infrastructure managers via Fire and Rescue Service Control.

8C9.37 This will ensure:

- communication is made with the appropriate, predetermined infrastructure representative

- messages are transmitted, received and recorded in an appropriate way

- any discrepancies or queries can be recorded and clarification sought through agreed communication routes.

8C9.38 Trackside telephones fall into three categories:

- **Signal post to signal box**
  These are identified by black and white hatching on the cabinet lid. They can be in groups of up to 12 on one circuit, however only one can be used at a time. Most multi-aspect coloured light signals incorporate a telephone, but only a small number of semaphore signals will have them.

- **Electrification telephones**
  These are identified by a red telephone symbol on a white background and provide direct communication with the electrical control room. These are specific telephones that cannot be used for any other purpose.

- **Other telephones**
  These are identified by a black diagonal cross on white background. They enable communications between two specific points, but may be used as dialling telephones giving access to the whole of the Network Rail systems.

PART C–10
Rail vehicles

General

8C10.1 There are many different types of rail vehicles in use on the infrastructure, and it is not the intention of this guidance to cover any one type in any detail. Each Fire and Rescue Service must ensure that suitable and sufficient training and familiarisation is available to staff on the variations of rail vehicles that may be operating within their area. This section provides some information on the main types of trains along with a number of key components and features of construction that may be relevant to Fire and Rescue Service staff when undertaking operations or training with rail vehicles.

Types of rail vehicles

8C10.2 There are a number of categories of trains within the United Kingdom providing services for passengers and freight. They are categorised as the following:

- multiple units

- locomotive hauled or propelled

- high speed trains

- Eurostar and inter-continental trains.

Multiple units

8C10.3 The majority of passenger trains are made up of multiple units and are used mainly for local services and suburban services. All vehicles in a unit are normally designed as passenger carriages but exceptions can occur. Gangways are provided to allow movement of passengers and crew between vehicles. The outermost vehicles will have driver's cabs. A number of units can be operated together as a single train. Some units are equipped with gangways at the cab ends to allow movement between different units when formed into a multi-unit train. Units without end gangways for regular operation in narrow tunnels have end doors to permit evacuation.

8C10.4 There are three types:

- Electric multiple unit (EMU)

- Diesel electrical multiple unit (DEMU)

- Diesel multiple unit (DMU) (mechanical or hydraulic transmission).

Electric multiple unit (EMU)

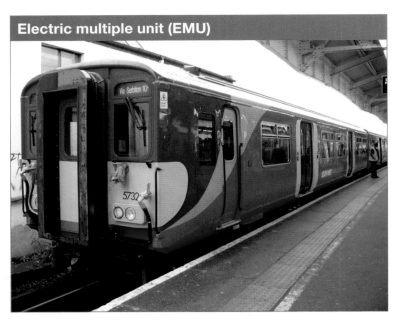

Diesel electric multiple unit (DEMU)

Diesel multiple unit (DMU)
(mechanical or hydraulic transmission)

Electrical multiple unit (EMU)

8C10.5 An electrical multiple unit is a multiple unit train powered solely by electricity. The electrical multiple unit can be subdivided further into three categories i.e. those that are powered by alternating current electric or direct current electric, or those that use dual voltage.

8C10.6 The significant difference between these categories of trains is that alternating current electrically powered trains are supplied by overhead line equipment and utilise a pantograph to connect with the power supply. Whereas direct current electrically powered trains use a third rail supply located at ground level. Dual voltage trains will have both sets of equipment to facilitate pick up of traction current from either alternating current overhead line equipment or direct current third rail.

8C10.7 Direct current powered trains use electrically interconnected 'pick up shoes' to connect with the power supply. The cars that form the electrical multiple unit set can usually be separated by function into these four types: Power car, motor car, driving car and trailer. Each car can have more than one function such as a motor driving car or power driving car.

8C10.8 An electrical multiple unit consists of two 'driving cars' situated at each end of the train. At either end of each car will be a drivers cab for controlling the train.

8C10.9 A 'trailer car' is any car that does not carry traction or power related equipment and is similar to a passenger car in a locomotive hauled train.

8C10.10 A 'power car' carries the necessary equipment to draw power from the electrified infrastructure, such as shoes for the third rail system and the pantograph for the overhead lines system and transformer.

8C10.11 A 'motor car' carries the traction motor. Alternating current electrical units use overhead line equipment whereas direct current electrical units use third rail supply. All of these units motors are located below the floor.

8C10.12 Rheostatic braking (sometimes called dynamic, or electrical braking) is also available on different multiple units. By attaching the motors to an electrical load they act as generators and slow the train down. The load is transferred to resistors which dissipates electrical energy as heat to the atmosphere, though it is sometimes used to heat the interiors. These resistors generate significant heat and have been known to cause fires in rail vehicles.

Diesel electrical multiple unit (DEMU)

8C10.13 In a diesel electrical multiple unit a diesel engine drives an alternator which produces alternating current current. This current can then be fed to electric traction motors in the wheels or the bogies.

8C10.14 In most modern diesel electrical multiple units each car is entirely self contained and has its own engine, generator and electric motors.

Diesel multiple unit (DMU) (mechanical or hydraulic transmission)

8C10.15 A diesel multiple unit is a multiple unit train powered by one or more on board diesel engines. It can fall into two categories based on their form of applying motive power to the wheels.

- **Diesel mechanical** – In a diesel mechanical multiple unit the engine power is transmitted via a gearbox and driveshaft directly to the wheels of the train, much like a car.

- **Diesel hydraulic** – In diesel hydraulic multiple units a hydraulic torque converter, (a type of fluid coupling), acts as the gearbox and may have an external oil cooler. The hydraulic unit is connected with a driveshaft in the same way as a diesel mechanical transmission.

Locomotive hauled or propelled

8C10.16 A locomotive has no pay load capacity of its own and its sole purpose is to move the train along the tracks. The locomotive or locomotives may be positioned at either end of the train or both and can be powered by diesel electric transmission or electric (alternating current). Locomotive hauled passenger trains have been largely replaced by multiple units but are still used on inter-city services and for special trains. Regular locomotive hauled trains are usually operated as if multiple units as fixed push-pull sets.

8C10.17 For push pull passenger operation a driving trailer is provided at the other end of the train which is an unpowered vehicle with a full driver's cab. The majority of driving trailers do not carry passengers.

High speed trains

8C10.18 This category is focused on trains that operate national high speed passenger services. Train design and construction has advanced considerably over recent years which has enabled train operating companies to operate faster and more comfortable services to their customers.

8C10.19 The Network Rail High Speed Train or 'HST' is operated as a multiple unit but consists of two single ended locomotives between which are coupled seven to nine trailer carriages'

Eurostar and inter-continental trains

8C10.20 The 'Class 92' has been specifically constructed to haul freight and passenger services on United Kingdom and European rail networks and operate through the channel tunnel. These locomotives are not significantly different from any others with the exception that they are dual-voltage. Eurostar trains are notable in that the units are very long (20 vehicles) and most of the vehicles share bogies – articulated. Due to channel tunnel requirements they are built to a very high standard. Power cars are located at the train ends (essentially single ended locomotives). The trailers are unpowered. Differences between United Kingdom locomotive units and European locomotive units are minimal. At present no other 'foreign' trains operate in Great Britian.

Key components

8C10.21 It is in the interest of every firefighter to familiarise themselves with the principal components of train construction. There are a number of key construction features that are consistent in most types of train that operate on the rail network, these are listed below:

Weight

8C10.22 The weight of rail vehicles can be significant. Main-line locomotives typically weigh between 80 and 126 tonnes, a trailer carriage between 30 and 45 tonnes, a multiple unit vehicle between 40 and 60 tonnes. Freight vehicles can range between 10 tonnes empty and 100 tonnes fully loaded depending on the size and type.

Air systems

8C10.23 Except in the case of heritage railways, trains are air braked and most passenger vehicles also have air suspensions. Single or dual brake pipes will connect between vehicles and all vehicles will have air reservoirs. Locomotives and some multiple unit vehicles will have compressors fitted.

Cant rail

8C10.24 The cant rail is represented as an orange safety line marked around the upper section of the entire train and represents the minimum safe distance from the overhead line equipment. If any part of the body does go beyond this line there is a risk of electrocution due to the high voltage electrical supply. Fire and Rescue Services should note that this rail is normally only found on passenger vehicles and locomotives, freight vehicles are not marked.

Pantograph

8C10.25 A 'pantograph' is a device that conducts electric current from the overhead line equipment to electrical powered trains or trams. The pantograph is spring loaded and pushes a contact head up against the contact wire to draw the electrical power needed to run the train. The steel rails on the tracks act as the electrical return.

Bogies

8C10.26 A bogie is a small truck to which the wheels are attached and can swivel to allow vehicles to negotiate curves easily and safely. Bogie vehicles will normally have two bogies, at each end, but in some cases a bogie may be shared. If in an accident a bogie becomes detached it can, due to its mass (typically between 4 to 8 tonnes each for a carriage), cause considerable damage.

Third rail shoe

8C10.27 Electricity is transmitted to the train by means of a 'contact' shoe or sliding 'pick-up' shoe. The shoe is normally in contact with the third rail.

8C10.28 Where there are gaps in the conductor rails there can be significant arcing as contact is broken and restored. It should be assumed that (for non-London Undergrounds Ltd trains) if the shoe at one end of a unit is in contact with the third rail that all shoes on all vehicles in the unit are live.

Buffers or stop

8C10.29 Normally rail vehicles are fitted with automatic couplings, however some may have buffers. A buffer is a shock absorber and can form part of the coupling system for railway vehicles. Where this is the case, each wagon, car or carriage will have two buffers at each end. They are bolted to the main frame of the vehicle and contain shock absorbing pads. When carriages are coupled together the buffers are brought into contact with those on the next vehicle. A screw coupling is used between each pair of vehicles to keep the buffers pressed together.

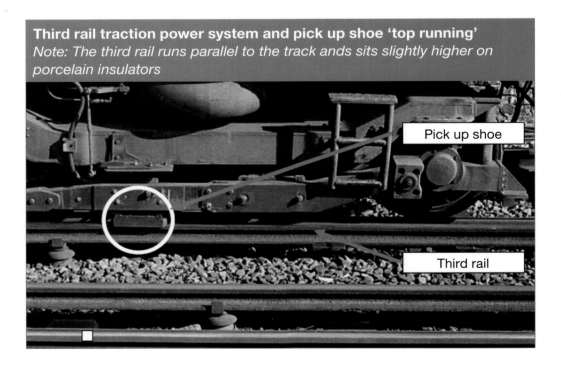

Third rail traction power system and pick up shoe 'top running'
Note: The third rail runs parallel to the track ands sits slightly higher on porcelain insulators

Pick up shoe

Third rail

Couplings

8C10.30 A 'coupling' is a mechanism for connecting railway cars, carriages and wagons in a train. They are a standard design and consist of a towing hook and chain.

Batteries

8C10.31 Batteries can be found on a wide range of rail vehicles including locomotives and passenger carriages. In addition to supporting traction batteries will also provide power to a range of services including air conditioning, cooling ventilation, lighting etc. Battery powered rail vehicles are also used for maintenance work on London Underground and on some industrial and heritage railways.

Construction materials

Older rolling stock

8C10.32 Older rolling stock is regularly used on different rail systems. Some of these have been subject to upgrades and modernisation, and will not present any more hazards than more modern stock. However some older stock, particularly those used on some heritage and pleasure systems may represent additional hazards due to the materials of construction. The Fire and Rescue Service should therefore ensure that they have a good awareness of the different types of rolling stock used in their area and any additional hazards they may present.

Asbestos

8C10.33 Asbestos was primarily used as an insulating material between the inner and outer skins of carriages. Most modern (post 1980) mainline rolling stock is unlikely to contain significant amounts of asbestos, however small amounts may still be present in some individual components.

8C10.34 As part of general refurbishment, older rolling stock is subject to a programme of asbestos removal from the accessible areas, but some may still be present in enclosed structural areas, however there are now very few vehicles that are pre-1980.

8C10.35 In most circumstances the likelihood will be minimal, as the overwhelming majority of stock no longer contains asbestos. However the likelihood is increased where private rail companies including heritage railways who operate their services on Network Rail track are using older rolling stock. Therefore, firefighters should remain vigilant to this risk.

8C10.36 If a passenger carriage is involved in an incident, the Incident Commander should check whether there is an asbestos hazard by contacting the railway control office quoting the carriage identification number (painted on the waist on each side of one end). Until it is confirmed that blue asbestos is not present, personnel should assume that it is and take the appropriate precautions.

Glass

8C10.37 In the interests of passenger safety, modern railway carriages are designed to contain passengers in emergency situations. This strategy includes the progressive fitting of laminated glass windows.

8C10.38 Passenger rail vehicles have large areas of glazing incorporated in their design, with modern glazing systems designed to prevent intrusion and contain passengers within the vehicle in the event of an incident.

8C10.39 Many of the vehicles are designed as sealed air conditioned environments, though a significant number of both modern and heritage vehicles are designed with opening lights to provide natural ventilation.

8C10.40 The glazing system will generally be based around a replaceable double glazed unit set within the vehicle body. The construction of the unit uses two types of glazing to provide the high level of performance and protection required.

8C10.41 The outer (weather side) element is toughened with the inner element (passenger side) element being laminated

8C10.42 It should be noted that the lamination in certain marked or signed windows, designed to provide emergency egress, will be resin based as opposed to plastic. This makes the glazing easier to break and provides more ready access to the passenger cell

8C10.43 In older or heritage stock the glazing system may not be as sophisticated being constructed from toughened glass.

Polymer composite materials

8C10.44 Polymer composites, also known as machine/man made mineral fibres (MMMF), are regularly used in the construction of rail vehicles.

8C10.45 Hazards from fire and impact damaged composites will be present in a variety of forms at accident sites and depending upon the volume and concentration of these hazards, can present a range of risks to those responding to the accident. Hazards can be presented in the following forms:

- toxicity

- particulates (fibres and dusts)

- conductivity

- change in structural strength (affects upon rescue operations)

- contamination.

8C10.46 More information regarding polymer composites can be found in the national operational guidance dealing with hazardous materials.

Construction features

Fire resistance

8C10.47 Rail vehicles of modern construction will deliver low levels of flammability due to the widespread use of fire resistant construction materials. Many modern rail vehicles have unique design features which can include safety installations such as automatic fire detection, upgraded fire separation, passenger communication systems, fire alarm call points and indicator panels. These features will assist in reducing the incidence of fire and fire development where it does occur. However, it should be recognised that for passenger rail vehicles a significant proportion of the fire load will be due to baggage and other items they bring aboard. The fire loading on freight vehicles can be very significant in terms of fire loading due to the variations and quantities of cargo carried.

Air conditioning systems

8C10.48 Trains are now regularly fitted with air conditioning units to stabilise the ambient temperature for the passenger. On the majority of modern vehicles the units are roof mounted as complete modules and can thus be easily identified. On older vehicles the equipment was underframe mounted. On passenger vehicles where the saloon windows can be opened or partially opened it can be assumed that air conditioning is not fitted.

8C10.49 Train air conditioning units are fitted with a liquefied refrigerant which is hazardous, as it is a corrosive. This liquid easily vaporises into a gas when exposed to any small temperature increase either by a rise in ambient environmental temperature or heat generated by the operation of the air conditioning unit. The vaporisation process can generate increasing pressure within the air conditioning system of up to 10 bars pressure. A serious hazard is then created if the unit is ruptured or cut during rescue operations.

High voltage connectors

8C10.50 High voltage couplings can be found on a number of trains and rolling stock. These were originally introduced to supply increased electrical voltage from carriage to carriage on trains designed for inter continental journeys (e.g. the Eurostar or freight trains travelling between the United Kingdom and the continent). Now these couplings can be found on a number of freight trains and other locomotive units used for pulling passenger rolling stock.

8C10.51 On multiple units inter-vehicle connectors are used to distribute traction power through the units. On some alternating current electrical multiple units there will be a 25 kiloVolt power line running along the roof with connections made between vehicles. On locomotive hauled stock power for electric train heating will be supplied from the locomotive using high voltage connectors. Other, low-voltage connectors may be encountered for control signals between vehicles. The overwhelming majority of freight vehicles do not have electrical connections to other vehicles or electrical systems at all.

Polychlorinated biphenyls (PCB)

8C10.52 Although polychlorinated biphenyls are no longer used in new electric locomotives and coaches of electric trains, older models may be encountered where these substances are still in use in transformers and capacitors.

8C10.53 All train operating companies and Network Rail are in the process of replacing all components containing. polychlorinated biphenyls.

8C10.54 Polychlorinated biphenyls are organic, oil soluble materials of moderate toxicity. The main risk to firefighters is via 'skin absorption' or inhalation. Although a number of toxicological effects can arise from prolonged or chronic exposure to substances containing polychlorinated biphenyls, firefighters are likely to experience short term exposure which would at the most result in irritation to the skin, eyes, nose and respiratory tract.

8C10.55 The presence of polychlorinated biphenyls in equipment will be indicated by a label which will be found on the equipment and the vehicle body.

Lifting points

8C10.56 All rail vehicles have lifting points that are used for maintenance, servicing and for re-railing purposes. These can be readily identified by an external marking.

Internal electrical systems

8C10.57 Increasing availability of electrical supplies to power on board computerised systems, catering facilities, air conditioning units, three pin electrical outlet sockets for customers to plug in mobile phones and computers require increased electrical generation and output, electrical components and cabling.

8C10.58 Therefore personnel need to be aware that creating access points into the bodywork of a train could be dangerous exposing electrical equipment and cabling that could range in voltage from 110 volts to 875 volts.

8C10.59 Electrical equipment including traction equipment, batteries, battery charges etc is normally housed in the underframe area, the majority of which is encased in a metal skirt. However electrical equipment can also be located on a carriage roof as in this example which includes air conditioning unit, bus lines, brake resistors and a pantograph (A).

8C10.60 In some designs of rail vehicles electrical cables run directly through the carriage from the pantograph located on the roof to the main transformer which is located in the underframe. The cable is shielded from the passengers by an earthed protective metallic casing which in turn is shielded by an aesthetic shroud (B).

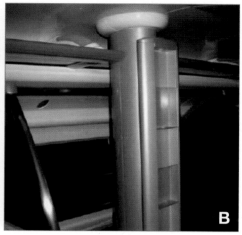

Toilets and sewage systems

8C10.61 Older trains may still deposit human waste along the tracks. These vehicles are progressively being phased out by replacement or refurbishment but will remain in significant numbers for the next 10-15 years. This can attract rats and the risk of Weil's disease (Leptospirosis) should not be overlooked.

8C10.62 All modern trains are designed and constructed with toilet retention systems. These tanks are generally not emptied for four to five days until the train is routinely serviced. This means that there can be the potential for over 400 litres of stored human waste, which could be accidentally released during an incident or a tank could rupture resulting in the risk of contaminating emergency service personnel working within a risk area.

8C10.63 The tanks may not always be in obvious locations. On some recent vehicles the tanks have been located inside the vehicle due to lack of space on the underframe (e.g. Pendolinos, Voyagers).

Detonators

8C10.64 Otherwise known as 'fog signals', these are small explosive devices used to warn train drivers to stop immediately as a major hazard is further up the track.

8C10.65 Originally stored in trackside cabins, now storage of the detonators is in more secure locations in buildings or in train cabs to avoid the risk of being stolen and misused. On trains, 12 detonators are required to be carried in each drivers cab. If misused they become an explosive and projectile hazard to firefighters.

Thermite

8C10.66 This material is used principally for jointing sections of track. The substance is carefully managed by infrastructure managers and unless being used at a rail worksite will be stored in depots and marked in accordance with relevant hazardous chemical signage. Additional hazard to Fire and Rescue Service crews associated with this substance is that when involved in fire it reacts violently to the application of water.

Sleeping and catering facilities

8C10.67 Some rail vehicles have on board catering facilities which may contain electrically operated catering equipment. It is normally possible to isolate the circuits that power catering areas via dedicated circuit breakers which can be located in marked electrical cubicles.

8C10.68 Some routes operate sleeper services and Fire and Rescue Service personnel should be aware that incidents involving these rail vehicles may present a range of additional operational challenges particularly when searching carriages. The sleeping accommodation is described as 'sleeping berths'. Cabins may be occupied by one or more people and where this is the case berths are arranged in a 'bunk' formation.

Access/Egress

8C10.69 The preferred route for the Fire and Rescue Service when gaining access to carriages in emergencies should always be via the doors. Although variations exist, most modern stock is fitted with door release mechanisms that are operable by Fire and Rescue Service crews from outside. These release mechanisms can be of mechanical or compressed air design, and can be operated by either door release handles or air cocks located in the lower body of carriages, which are labelled.

8C10.70 Where passenger access doors do not provide safe or rapid access either due to damage or position two further options can be considered:-

The corridor connector

8C10.71 The corridor connector is designed to provide a weather proof connection between two rail vehicles. Depending on its design it may be feasible to cut through the connector and access the vehicle end doors. This technique may not be viable due to the number of jumper cables that may be found in this area on some vehicles

Windows

8C10.72 Whilst the glazing systems in modern rail stock are designed to survive severe impacts they can be defeated using standard Fire and Rescue Service equipment and techniques. The outer, toughened, element can be 'managed' using techniques such as a glass hammer, punch or other patent glass management tool as would be used to manage road vehicle glazing. The inner, laminated, element will require significant effort to initially penetrate and then, with some form of tooth edged blade or cutting tool, to cut through the polyvinyl butyral lamination material to flap or remove the glazing. Whilst the laminated element is similar in construction to road vehicle laminated glass, the plastic membrane is significantly tougher. Where there are windows with purpose designed emergency egress these may utilise a resin based rather than polyvinyl butyral based lamination and will provide reduced resistance to rescue efforts. These windows are generally marked both inside and outside the vehicle.

8C10.73 Access/egress via windows would normally be a last resort for Fire and Rescue Service operations such as the removal of stretcher bound casualties where manoeuvrability of a stretcher through a damaged vehicle would increase risk.

8C10.74 Where large openings for access and egress are required, consideration may be given to cutting down the vehicle side directly under the side window. This will be dependant on a clear understanding of the vehicle construction techniques used.

8C10.75 Where carriages are fitted with double glazed window units the external pane will ordinarily be toughened glass and is likely to shatter during collision. The inside pane will normally be of laminated design and can be readily cut by using existing standard Fire and Rescue Service equipment.

8C10.76 Rolling stock consists of a wide variety of age and design of vehicles and the Fire and Rescue Service should seek to ensure that crews are familiar with the more widely found examples and any specific variations that may occur locally. This will greatly assist when planning for effective access to and from carriages in emergency situations.

Use of rail vehicles

Passenger rail vehicles

8C10.77 Passenger rail vehicles can vary greatly in size and complexity from small two carriage diesel powered units on local lines, to electrified Inter-City or high speed Eurostar units made up of many carriages which can include sleeper services. Local Fire and Rescue Services are advised to make regular visits to local rail operators to become familiar with the variations that they may be likely to face in emergency situations.

8C10.78 It is not possible to provide detailed layout diagrams of all passenger vehicle types in this guidance, however below are some generic diagrams that show some of the likely internal carriage configurations.

8C10.79 Further information regarding specific rail operators and rolling stock can be found by researching the list of websites and reference documents listed in Section 12.

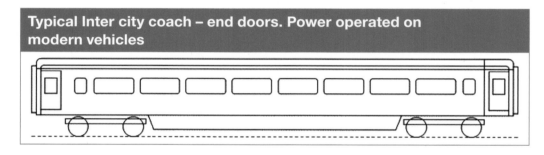

Typical Inter city coach – end doors. Power operated on modern vehicles

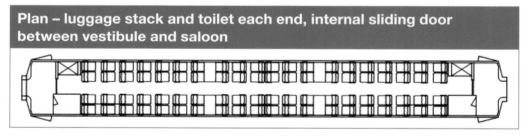

Plan – luggage stack and toilet each end, internal sliding door between vestibule and saloon

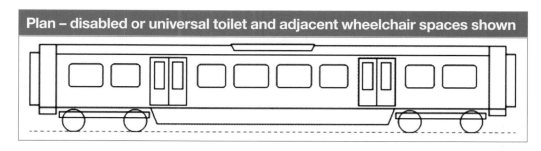

Plan – disabled or universal toilet and adjacent wheelchair spaces shown

Plan – luggage stack and toilet each end, internal sliding door between vestibule and saloon

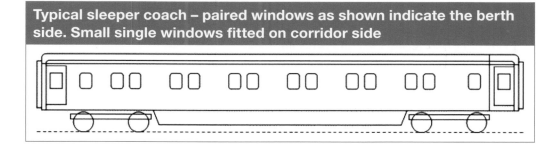

Typical sleeper coach – paired windows as shown indicate the berth side. Small single windows fitted on corridor side

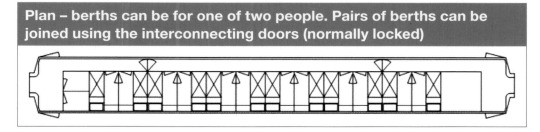

Plan – berths can be for one of two people. Pairs of berths can be joined using the interconnecting doors (normally locked)

Freight rail vehicles

NON-HAZARDOUS FREIGHT

8C10.80 Everyday non-hazardous goods such as food, drink and clothes as well as steel products from ore to finished items, construction products, including aggregates, cement and blocks are carried by rail, with a significant amount of the freight operation being undertaken at night. Over 800,000 deep sea containers move inland by rail per year from major ports including Felixstowe and Southampton, and there are regular services, including temperature controlled ones, to and from the continent through the Channel Tunnel.

HAZARDOUS FREIGHT

8C10.81 The conveyance of hazardous substances, by rail, occurs on a frequent basis throughout all areas of the country. The movement of such materials and substances can be at any time during the day or night. All personnel need to be mindful that any incident involving a train could potentially involve a multitude of hazardous substances.

8C10.82 Legislation and marking systems for the transportation of hazardous substances apply predominantly to both road and rail. This is covered in some detail in the national guidance for dealing with incidents involving hazardous materials, and this should be read in conjunction with this section.

Total operating processing system

8C10.83 Train operating processing system is a computerised system enabling the National Rail Network to keep a constant check on the position and availability of every rail vehicle on the national rail system and provide specific information of the various loads that are being hauled.

8C10.84 It consists of a central computer system connected to regional control offices, marshalling yards and depots throughout the country. The freight operating companies are responsible for keeping the information up to date.

8C10.85 The freight operating company will feed into the computer all the wagon and freight details, loaded or unloaded, freight train movements and type of traffic conveyed. An Incident Commander via Fire Service Control can obtain any specific information from the system, on any wagon or freight train and its cargo.

8C10.86 Each wagon is clearly marked with an individual identification number on the side of each wagon. This number will be recorded on the train operating processing system computer to enable information on the load to be readily made available for operational personnel in the event of an incident.

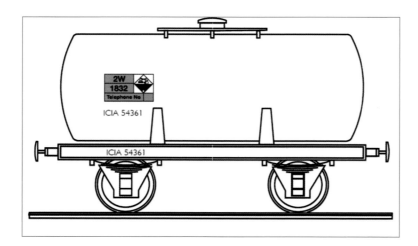

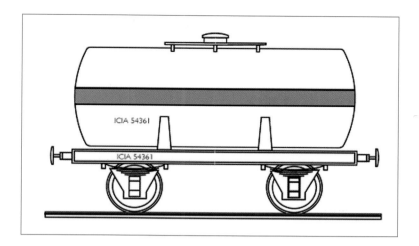

Additional tank wagon identification schemes

8C10.87 Tank wagons carrying certain classes of dangerous goods can be identified by specific colour schemes. For example, tank wagons carrying liquid petroleum gas will have a white barrel with a horizontal orange stripe round the barrel at mid height. Tank wagons carrying flammable liquids are painted dove grey and the sole bars are painted signal red. (Sole bars are the two horizontal metal bars upon which the bottom of the tank rests.)

8C10.88 In addition to the labelling, all wagons containing dangerous goods will have a 'Network Rail Dangerous Wagon Label' displayed on each side. This label indicates the class of substance being carried and the principal hazard encountered. Containers that are hauled by rail are exempt from being labelled in this manner.

Information available from the rail vehicle crew

8C10.89 Information on the specific details of each wagon and the details of their individual loads is held by the rail vehicle crew. This information sheet is known as the 'consist'. Information found on the 'consist' will contain the following information:

- position of every wagon from front to rear
- train identity number
- locomotive identity number
- wagon numbers
- dangerous goods emergency codes
- United Nations number plus specialist advice contact code.

8C10.90 It would be advisable for the Incident Commander, to obtain the consist as this will provide immediate information on the type of loads being carried and provide interim guidance on a course of action to follow until further information is secured either from the total operating processing system or from a designated specialist adviser.

Guidance manual for accepted and regulated materials to be transported

8C10.91 Network Rail has established conditions under which they are prepared to accept dangerous goods detailed in their *'List of Dangerous Goods and Conditions of Acceptance'*. This is a working manual for all Network Rail staff commonly known as the 'pink pages' which sets out the specific instructions to staff on the handling of dangerous materials.

8C10.92 It contains guidance on the action necessary in emergencies, explains how to obtain specialist assistance, and requires staff to summon emergency assistance when an incident occurs. It also includes illustrations of the labels used on packages of dangerous substances and on the wagons containing them.

8C10.93 All train crews should be aware of all goods carried, especially those of a hazardous nature and should ensure that compliance with the regulations is made, in respect to the quantity, labelling and method of transportation and to initiate emergency measures if required.

Conveyance of explosives by rail

WAGON LABELLING

8C10.94 The wagon label for explosives (both commercially and military) replace the individual UN hazardous substance number with a set of four characters identifying the category of the explosive. An example of this is a wagon label whose emergency code begins with the characters 1.1D would indicate an explosive of Hazard Division 1.1, compatibility group D, (i.e. an explosive presenting a mass explosion hazard).

WEIGHT LIMITATIONS

8C10.95 Only commercial explosives are covered by GO/RT3422, (*'Railway Group Standard which identifies the requirements for the acceptance and carriage by rail of explosives on NR* and the total amount to be conveyed on any one train is limited to 36.25 tonnes.

8C10.96 Military explosives are covered by the Conveyance by Rail of Military Explosives Regulations 1977. The weight limits for military explosives, and for mixed commercial and military loads are given in GO/RT3053.

8C10.97 To summarise no more than 20 tonnes of explosives per wagon or container can be carried with a net amount of the same explosives. It is permissible to convey several groups of explosives on the same train but subject to a separation distance of 80 metres and a limitation of 40 tonnes per group. This is applicable to groups within Hazard Division 1.1 and combined with 1.3 and/or 1.5. For groups within 1.3 and 1.5 either separately or together the separation distance reduces to 40 metres and the limit per group increases to 120 tonnes.

WAGONS HELD IN SIDINGS

8C10.98 Railway wagons laden with explosives in transit are occasionally parked overnight in railway sidings and to this effect the working manual for rail staff is quite specific in detailing the responsibilities to staff where explosives are concerned. The manual places personal responsibility upon the supervisor for ensuring that any wagon of explosives standing in the sidings, goods or marshalling yards are under surveillance by railway staff and for keeping a tally of each wagon with its location whilst in their responsibility.

8C10.99 If through exceptional circumstances it becomes necessary to hold explosives in an 'unsupervised location' the Chief Operating Officer of Network Rail has made arrangements to ensure that Network Rail staff responsible for the handling and conveyance of dangerous goods will inform the British Transport Police and the Fire and Rescue Service whenever it is known that wagons containing explosives are likely to be held in unsupervised goods yards or sidings.

8C10.100 On receipt of this information the Fire and Rescue Service concerned should instruct a Fire Safety Officer to visit the site and if conditions warrant it, arrange with their Service Control for an appropriate special attendance to be made to the sidings in the event of any fire or special service call being received.

Irradiated fuel flasks

8C10.101 If an incident occurs involving an irradiated fuel flask, National Arrangements for Incidents Involving Radioactivity (NAIR) will be activated immediately. However, there will be some operational priorities to be considered whilst the National Arrangements for Incidents Involving Radioactivity (NAIR) scheme is instigated.

8C10.102 In the unlikely event that an irradiated transport flask is subjected to structural damage initial Fire and Rescue Service responders should keep at least 50 metres away upwind and uphill of the incident. If this area has to be entered for rescues, firefighters should wear appropriate protective equipment in line with radiation procedures.

8C10.103 Under severe damage to a flask there could be an immediate danger of a release of radioactivity and the approach to the incident and the immediate area around the incident should be treated with great caution and cordoned off to create safe areas.

8C10.104 If water leaks from the flask always treat as if radioactive and avoid if possible, especially any contact with the eyes or skin. In the event of the flask being involved in fire, it should be cooled by water spray. The nuclear physicist will monitor the run off and inform the local water authority of the slight chance of radioactive contamination in water courses.

8C10.105 The risk of contamination is not limited to water sources as there is a high possibility contamination can occur through airborne radioactive particles. Monitoring of the wind speed and direction is vital to continue to stay upwind and avoid further contamination risks.

Maintenance vehicles

8C10.106 A wide variety of specialist service and maintenance vehicles are deployed on the rail networks. These vehicles are specifically designed to carry out specific tasks which may include, 'steaming' of the track, to remove leaves and de-grease the rail, or technical equipment for examination of the rail line.

8C10.107 These vehicles can vary widely in terms of size and traction systems from relatively small road-rail vehicles powered by diesel to complete high speed inspection trains powered by electricity.

8C10.108 These vehicles can be fitted with a wide range of specialist equipment designed to support the function of the vehicle and may include items such as, quantities of chemicals, and various scientific apparatus, including isotope equipment.

PART C–11
Specialist equipment

Fire and Rescue Service

Local

8C11.1 Local Fire and Rescue Services will provide a range of different pieces of equipment designed to assist operations at railway incidents. These will vary according to local assessment of risk and local procurement routes. Some Fire and Rescue Services have introduced particular specialist vehicles into service to facilitate access for crews and equipment in certain challenging local circumstances, some examples are the London Fire Brigade motorised rescue trolley and the Avon Fire and Rescue Service yellow peril road rail vehicle.

8C11.2 It would be helpful to Fire and Rescue Services for firefighters and managers to consider the types of equipment available in neighbouring Fire and Rescue Services, within their region and in neighbouring regions that may form part of any intervention strategies when determining local plans.

National

8C11.3 Vehicles and equipment provided as part of national resilience such as certain urban search and rescue modules may be able to provide additional capacity and capabilities which may be required at major railway incidents.

Non-Fire and Rescue Service

8C11.4 The rail industry will be able to provide a range of specialist equipment to support Fire and Rescue Service operations. Access to this equipment would normally be through the on-site representative. Some examples of the type of equipment that may be available are as follows:

- Rail lighting
- Lifting equipment
- Command support
- Road/rail vehicle

- Short circuiting devices, these are normally deployed by personnel under the control of the Infrastructure Manager or the train operating company. However specialist appliances in some Fire and Rescue Services carry short circuiting devices to support locally agreed operational procedures.

Specialist personnel

Responsible Person at Silver

8C12.1 The infrastructure manager will assess the nature of the incident and determine if it is appropriate to send a management representative with knowledge, experience, training and authority to the scene. The Responsible Person at Silver will not normally be on scene and may have to travel some distance, and may be delayed by traffic or cordon controls.

8C12.2 The responsibilities that the Responsible Person at Silver may undertake include:

- representing the rail infrastructure manager at Silver level
- assess the suitability of control measures implemented
- implement and advise on additional rail specific control measures
- provide information to the Incident Commander on the impact of the incident to the wider infrastructure, to assist the Incident Commander in reviewing the incident plan
- provide suggestions on rail specific considerations to reduce the impact of the incident on the infrastructure or travelling public
- arrange for the attendance of rail specialist personnel to facilitate early return to normality, for example building engineers to inspect railway bridges or arches
- arrange for specialist equipment to be delivered, through the Incident Commander and relevant police service
- co-ordinate the phased reopening of rail lines as appropriate.

8C12.3 Different rail infrastructures have assigned the Responsible Person at Silver with different titles.

Network Rail

8C12.4 Network Rail may despatch a Rail Incident Officer to perform the role of Responsible Person at Silver. Normally an estimated time of arrival will be provided.

8C12.5 The Rail incident Officer is the nominated and certified person who is responsible for on-site command and control of all rail related operations. Not normally on-site before the arrival of the Fire and Rescue Services, they will be nominated to attend incidents that are likely to have a serious impact on the rail network. This Responsible Person will provide silver level liaison at scene for Network Rail infrastructure. This person is a railway expert and will be able to:

- provide information to the Incident Commander on rail safety matters

- arrange for specialist rail workers, engineers, contractors and equipment to be moved to the scene

- obtain specialist information on the infrastructure

- arrange the delivery of extra equipment such as rail trolleys, generators, cranes, welfare facilities.

- liaise with the Incident Commander to provide options for partial or complete restoration of service.

London Underground

8C12.6   The Responsible Person at Silver is referred to as 'Silver'. This person will normally be on-site at underground stations, or will travel to the scene of a surface incident. They can assist the Fire and Rescue Service in implementing appropriate controls and the attendance of rail industry specialists. They can also arrange:

- a specially equipped and trained team to assist jacking and moving rail vehicles, water removal and body recovery

- drinking water for passengers

- welfare facilities for post incident customer support

- additional rail staff

- provision of dedicated London Underground Ltd command vehicle.

Rail Accident Investigation Branch (RAIB)

8C12.7   The Rail Accident Investigation Branch (RAIB) is the independent railway accident investigation organization for the United Kingdom. It investigates railway accidents and incidents on the United Kingdom's railways to improve safety:

8C12.8   The Rail Accident Investigation Branch (RAIB) must by law investigate all rail accidents involving a derailment or collision which result in, or could result in:

- the death of at least one person

- serious injury to five or more people; or

- extensive damage to rolling stock, the infra-structure or the environment.

8C12.9 The Rail Accident Investigation Branch (RAIB) may also investigate other incidents which have implications for railway safety, including those which under slightly different circumstances may have led to an accident.

8C12.10 If an accident or incident occurs, the Rail Accident Investigation Branch (RAIB) will decide its response, which will be influenced by:

- whether the investigation is mandated by law

- whether there is important evidence at the scene

- whether it is part of a trend; or

- the safety issues at stake.

8C12.11 To carry out its investigations the Rail Accident Investigation Branch (RAIB) has appointed and trained Inspectors recruited from the railway industry and other investigating bodies. They are experienced, have a broad mixture of skills across the railway industry and have been trained in investigation techniques. All inspectors carry an Rail Accident Investigation Branch (RAIB) warrant card, which identifies their powers at the scene of an investigation.

8C12.12 The Rail Accident Investigation Branch (RAIB) can appoint others to assist them in an investigation, including Accredited Agents, who may assist in recording evidence at the accident site, and specialist services.

8C12.13 The Rail Accident Investigation Branch (RAIB) Inspectors have the power to:

- enter railway property, land or vehicles

- seize anything relating to the accident and make records

- require access to and disclosure of records and information; and

- require people to answer questions and provide information about anything relevant to the investigation

- a protocol between Rail Accident Investigation Branch and the Fire and Rescue Service is in development.

British Transport Police

8C12.14 The British Transport Police are the national police force for railways. They provide a policing service to the rail operators, their staff and rail users throughout England, Wales and Scotland. There are a few exceptions, eg. The British Transport Police do not police heritage railways. British Transport

Police officers carry the authority and responsibility of the Home Office Police on the rail network, however they are also trained in track safety which Home Office Police are not.

8C12.15 There is a Memorandum of Understanding between Rail Accident Investigation Branch, Health and Safety Executive and British Transport Police in respect of investigations at railway incidents. During initial investigations it will be a joint investigation pending the establishment of whether a crime has been committed. (The Memorandum of Understanding is on the Rail Accident Investigation Branch website).

8C12.16 At an incident British Transport Police or the Home Office Police Service in attendance will undertake cordon control and other police responsibilities.

Office of Rail Regulation

8C12.17 The Office of Rail Regulation is the combined safety and economic regulator for the railway and is responsible for the enforcement of health and safety legislation on the railway. Office of Rail Regulation will normally carry out its investigations in conjunction with British Transport Police and Rail Accident Investigation Board.

8C12.18 Office of Rail Regulation will normally investigate all significant incidents on the railway where workers or passengers are killed or seriously injured in liaison with British Transport Police and Rail Accident Investigation Board.

8C12.19 Multiple unit consists of two 'driving cars' situated at each end of the train. At either end of each car will be a drivers cab for controlling the train.

Section 9

Appendices

Example of Handover Form – Closing The Incident

Incident Handover Form

Incident Identification

Type of Incident ....

Incident No						

Location/Address/Map reference

....

....

....

....

....

....

Date of call

Day		Month		Year			

Time of call

Hours		Minutes	

Originator ....

The incident is now left in your sole charge and control. The following instructions/guidance is provided for your sole safety and for the safety of others.

....

....

....

....

....

Continue on additional numbered sheet if required.

Person Receiving/Accepting Control of Site

Name (print)..

Role...

Signature..

Date

Day		Month		Year			

Time

Hours		Minutes	

Witness

Name (print)..

Role...

Signature..

Date

Day		Month		Year			

Time

Hours		Minutes	

Section 10
Acknowledgements

The Chief Fire and Rescue Adviser is indebted to all those who have contributed their time and expertise towards the production of this operational guidance manual. There were many consultees and contributors, but the following deserve special mention.

Strategic Management Board

Phil Abraham, Fire Service College

John Barton, Retained Firefighters Union

Chief Fire Officer Chris Griffin, Association of Principal Fire Officers

Assistant Chief Fire Officer Peter Hazeldine, Chief Fire Officers Association

James Holland, Network Rail

Bob Ibell, British Tunnelling Society

Deputy Chief Fire Officer John Mills, Chief Fire Officers Association

David Morison, Devolved Administration Scotland

Jenny Morris, Health and Safety Executive

Peter Moss, Fire Brigades Union

Neil Orbell, London Fire Brigade

Peter O Reilly, Devolved Administration Northern Ireland

Andrew Sargent, Devolved Administration Wales

Craig Thomson, Fire Officers Association

Chris Waters, Institute of Fire Engineers

Core Delivery Group

Sarah Allen, Avon Fire and Rescue Service

Eddie Ball, West Yorkshire Fire and Rescue Service

David Bulbrook, London Fire Brigade

Andy Cashmore, West Midlands Fire and Rescue Service

Simon Dedman, Essex Fire and Rescue Service

Jes Eckford Strathclyde Fire and Rescue Service

Graham Gardner, Northumberland Fire and Rescue Service

Graham Gash, Kent Fire and Rescue Service

Rick Hanratty, Avon Fire and Rescue Service

David Lovett, Derbyshire Fire and Rescue Service

Rob Mackie, Derbyshire Fire and Rescue Service

Chris Noakes Essex Fire and Rescue Service

Robbie Roberts, Avon Fire and Rescue Service

Paul Tod, Greater Manchester Fire and Rescue Service

Danny Ward, West Midlands Fire and Rescue Service

Andy Wood, West Yorkshire Fire and Rescue Service

Industry experts

Nick Agnew, Transport for London

Andy Barr, Transport for London

Jim Holland, Network Rail

Richard Davies, Network Rail

Anson Jack, Rail Safety Standards Board

Kevin Lydford, Network Rail

Richard Newell, Rail Safety Standards Board

Michael Tarran, Heritage Railway Association

David Walmsley, Confederation of Passenger Transport UK

Organisations

Avon Fire & Rescue Service

London Fire Brigade

Network Rail

Rail Safety Standards Board

Special thanks

Michele Kunneke, Project Support Coordinator

Photographs and illustrations

Thanks to the following individuals, organisations and FRSs for permission to use photographs and diagrams:

Avon Fire & Rescue Service

Greater Manchester Fire & Rescue Service

London Fire Brigade

Network Rail

Rail Accident Investigation Branch

Railway Safety & Standards Board

Railway Technical Web Pages

Hübner (Germany)

Section 11

Abbreviations and glossary of terms

A

ACPO – Association of Chief Police Officers

Alternating Current (AC) – Rail vehicle traction current operating at 25KV. Normally found on National rail infrastructure, delivered via OLE.

Air turbulence – A moving rail vehicle can create areas of positive and negative pressure. When moving at high speeds these pressures can cause equipment to move and people to fall

Authorised walking routes – Areas constructed to allow people to escape to a place of safety. When no facilities are provided the Incident Commander will determine the safest route, providing additional controls as appropriate.

B

Ballast – Crushed stone, nominally 48mm in size and of prescribed angularity, used to support sleepers, timbers or bearers both vertically and laterally

Bi-Directional Line – A line on which the signalling allows for rail vehicles to operate in both directions

Bogies – The wheeled, supporting structures on which a train runs.

Bridge strikes – Collisions where bridges are struck possibly affecting the stability of the structure of bridges

Bridging – When a rail vehicle re-energises an isolated section of track by bridging the gap between a live and an isolated section of track.

BTP – British Transport Police, a national police force responsible for policing on the majority of rail networks and some tram systems, including the London Underground network

Buffers or stop – A buffer is a shock absorber and it forms part of the coupling system for railway vehicles

C

Catenary – Originally the term used to denote an overhead power line support wire derived from the curve a suspended wire naturally assumes under the force of gravity. Now adopted to mean the whole overhead line system

CBRNE – CBRNE is an acronym for an event type caused by Chemical, Biological, Radiological, Nuclear or Explosive materials.

CCA – Civil Contingencies Act

CCTV – Closed Circuit Television

Cess – The part of the track bed outside the ballast shoulder that is deliberately maintained lower that the sleeper bottom. The area immediately outside the ballast shoulder but not between tracks

CFOA – Chief Fire Officers Association

Chief Fire Rescue Advisor (CFRA) – Provides strategic advice and guidance to ministers, civil servants, Fire and Rescue Authorities in England and other stakeholders (including the devolved administrations), on the structure, organisation and performance of the Fire Rescue Service.

Critical National Infrastructure (CNI) – A term used by governments to describe assets that are essential for the functioning of society and economy.

Couplings – A coupling is a mechanism for connecting rail vehicles cars, carriages and wagons.

Cordon control – Cordons are employed as an effective method of controlling resources and maintaining safety on the incident ground.

Cutting – An area excavated to permit a railway to maintain its level and gradient through high ground without excessive deviation from a straight course

D

Department for Communities and Local Government (DCLG) – Government department whose remit includes fire and resilience

DCOL – Dear Chief Officer Letter

DEMU – Diesel electrical multiple unit

Detection Identification Monitoring Teams (DIM) – Specialist teams used at hazardous material incidents.

DLR – Docklands Light Railway

DMU – Diesel Multiple Unit – the generic term for a diesel powered train where a separate locomotive is not required because the traction system is contained under various cars in the train.

Dangerous good – Any product, substance or organism included by its nature of by the regulation in any of the nine United Nations classifications of hazardous materials.

Detonator – A small explosive device clipped to a running rail by a railway employee to warn approaching train driver of a hazard ahead

'Down line' – A rail industry term for indicating the direction of rail vehicle movement. FRS personnel should exercise care when such terminology is used and ensure the direction of rail vehicle travel is confirmed.

E

Emergency Possession – A working area designated by the Incident Commander and implemented by the infrastructure manager. The purpose of an emergency possession is to secure public and first responder safety, when attending a rail incident. This is a temporary measure until a responsible person at silver can confirm that all appropriate safety measures are in place. This area will normally form the inner cordon.

Embankment – A filled area to permit a railway to maintain a safe and functional gradient across low ground without excessive deviation from a straight course.

EIA – Equality Impact Assessment

EMU – Electric multiple unit, an electrically powered train comprising two or more cars that can be driven and controlled as a single unit from the leading driving cab

EP – Evacuation Point, a location where people can move from a hazard area to a place of relative safety. Here they can await a evacuation train or move to a final exit

ERL – Emergency Response Locations, a section of the rail infrastructure where a vehicle is designed to stop in an emergency, and where the facilities are provided to allow firefighting intervention and passenger evacuation.

F

Fastening – Fastening hold the running rails to the sleepers. Collective name for bolts, washers and lock nuts used to secure the stretcher bar components.

FOCs – Freight Operating Companies, The companies who transport goods, but not passengers, on the national rail network.

Fourth rail – Occasionally a fourth rail between the running rails may be found which acts as a return circuit which may carry a current of up to 250 volts DC.

Four-foot – The area between the two running rails of a standard gauge railway.

Fire and Rescue Service (FRS) – Local Authority Fire and Rescue Service

FRA – Fire Rescue Authorities

FSC – Fire Service Circular

G

Generic Risk Assessment (GRA) – This is a document that details the assessment of hazards, risks and control measures that relate to any incident attended by FRS.

Generic Standard Operating Procedure (GSOP) – Is a list of possible operational actions and possible operational considerations viewed against the 'Managing Incident – Decision Making Model' which has been divided into the six phases of an incident.

Gold/Silver/Bronze Command – The standard management framework employed at complex or major incidents, mandated by the Civil Contingencies Act (2004)

H

H&S – Health and Safety

HSE – Health and Safety Executive. The government body responsible for protecting people against risks to health and safety arising out of work activities in Great Britain.

Her Majesty's Railway Inspectorate (HMRI) – See ORR (Office of the Rail Regulation)

Hazard – A hazard is anything that may cause harm.

Hazardous Area Response Team (HART) – Ambulance specialist response teams trained and equipped to work within the inner cordon of an incident.

Hazardous Material (HazMat) – Are referred to as dangerous /hazardous substances or goods, solids, liquids, or gases that can harm people, other living organisms, property, or the environment.

I

IALO – Inter Agency Liaison Officer

IM – Infrastructure Managers. The Organisation responsible for the management and maintenance of the infrastructure

IPs – Intervention Points. A location designated for the use of the FRS and provided with facilities to ease an emergency intervention

Incident Command System (ICS) – The nationally accepted incident command system, as detailed in *Fire and Rescue Manual Fire Service Operations Volume 2 – Incident Command*.

Incident Commander (IC) – The nominated competent officer having overall responsibility for incident tactical plan and resource management.

Inner Cordon – The Inner Cordon surrounds the area where potentially hazardous activity may be conducted and encompasses both the Hot and Warm Zones.

Integrated Risk Management Plan (IRMP) – This is the FRS published assessment of risk within their county/metropolitan boundaries and subsequent action plan to address these risks.

Isolation – The disconnection of electrical supply from a section of track.

L

LALO – Local Authority Liaison Officer

Level crossing – A level crossing can be described as a railway line crossed by a road or right of way without the use of a tunnel or bridge

Level Crossing – Unprotected (Passive) Crossings – These crossings have no warning system to indicate a train's approach.

Level Crossing – Protective (Active) Crossings – These crossings give warning of a train's approach to vehicle users and pedestrians through closure of gates or barriers, or by warning lights and/or sound.

Line Blocked – All trains are stopped with no traffic passing over that section of line

Lineside – Within the boundary fence to a distance no closer than 3metres to the nearest line

Lookout – A person employed by the Infrastructure Manager, who has been assessed as competent to watch for and to give an appropriate warning of approaching trains when people are working on a railway in operation.

LPG – Liquefied Petroleum Gas, essentially either propane or butane

LRV'S – Light Rail Vehicles

LUL – London Underground Limited

Local Authority (LA) – Local government body in a specific area that has the responsibility for providing local facilities and services, e.g. County or District Council.

M

Major incident – A major incident is any emergency that requires the implementation of special arrangements by one or more of the emergency services.

Metro – The term used to denote an urban railway, often partly underground, carrying large numbers of passengers on rail vehicles.

MoF – Manual of Firemanship

Motor car – Carries the traction motor

MoU – Memorandum of Understanding

N

NAIR – National Arrangements for Incidents involving Radioactivity

NEPACC – National Emergency Planning Co-ordination Committee

NR – Network Rail.

NICS – National Incident Command System

O

OLE or OHLE – Overhead Line Equipment, an assembly of metal conductor wires, insulating devices and support structures used to bring a traction supply current to suitably equipped traction units.

On or near the Line – The presence of personnel or equipment, within three metres of the track or electrified equipment with the potential for harm to people or property.

ORR – Office of Rail Regulation, the independent combined safety and economic regulator for the railways incorporating HM Inspectors of Railways.

Outer Cordon – The Outer Cordon designates the controlled area into which unauthorised access is not permitted.

P

Pantograph – Equipment conducting electric current from the overhead line equipment to provide traction current and service power (Air conditioning, lighting etc) to rail vehicles.

PCB – Polychlorinated biphenyls

Personal Protective Equipment (PPE) – Provided personal protective equipment issued by FRS, includes fire kit, boots, gloves etc.

Pick up shoe – Electricity is transmitted to the train by means of a 'contact' shoe or sliding 'pickup shoe'. The shoe is in contact with the third/fourth rail.

Polymer composites materials – General term to describe all composite materials used in construction.

Possession – See Emergency Possession

Predetermined Attendance (PDA) – The pre-planned FRS response to accidents/incidents.

Points – A mechanical device that enables trains to be guided from one track to another.

Power car – Carries the necessary equipment to draw power from the electrified infrastructure, such as shoes for the third rail system and the pantograph for the overhead lines system and transformer to provide traction current and power for services.

PSA – Passenger Services Assistant. An example of rail staff on a driverless vehicle system, who would provide the Responsible Person at Silver function.

PVB – Polyvinyl Butyral

Q

QDR – Qualitative Design Review

R

Reasonably practicable – To carry out a duty 'as far as reasonably practicable' means that the degree of risk in a particular activity or environment can be balanced against the likely success, time, trouble, cost and physical difficulty of taking measures to avoid risk.

Risk – Risk is the probability that somebody could be harmed by a hazard or hazards, together with an indication of how serious the harm could be.

Risk Assessment (RA) – A risk assessment is a careful examination of what, in the workplace, could cause harm to people, in order to weigh up whether enough precautions have been taken or more should be done to prevent harm. The law does not expect the elimination of all risk, but the protection of people, including those requiring assistance or rescue, as far as is 'reasonably practicable'.

Rail Accident Investigation Branch (RAIB) – is part of the Department for Transport, and is the independent railway accident investigation organisation for the UK.

Rail Safety Standards Board (RSSB) – An independent rail industry body which mangers the creation and revision of certain mandatory and technical standards (including Railway Group Standards) as well as leading a programme of research and development on behalf of government and railway industry.

Rail infrastructure – A general term that encompasses, traction systems, signalling equipment, fixed communication systems and all aspects of the built rail environment including; stations, termini, bridges, viaducts etc.

Rail system – A network, for example London Underground, Network Rail, Romney Hythe and Dymchurch Narrow gauge railway.

Rail Vehicle – Includes trams, rail carriages, trolleys, service vehicles. Regardless of size, length or power supply.

Responsible Person at Silver – A representative of a Rail System who has the authority, knowledge, training and experience to provide liaison and advise to the FRS Incident Commander at 'silver' level, providing information and options to support the overall plan.

Residual potential – The voltage left in electrification equipment that has been isolated but not earthed.

RIO – Rail Incident Officer. The responsible person at silver for Network Rail infrastructure. The RIO is a nominated and certificated member of Network Rail staff, who may respond to an incident to co-ordinate the railway response The principal contact point for the emergency services and train operating companies to advise on safe systems of work.

Risk – Risk is the probability that somebody could be harmed by a hazard or hazards, together with an indication of how serious the harm could be.

RPE – Respiratory Protective Equipment, provided for the protection of FRS personnel's respiratory system.

Run at caution – Train drivers are informed that there is an incident on or near the railway. The driver will reduce their speed in the area to ensure stopping safely.

Running Rails – The rails that the vehicle's wheels are guided along.

Rendezvous Points (RVP) – Pre-planned locations for the holding/gathering of emergency services when attending incidents on railway infrastructure.

S

Safe system of work (SSoW) – A procedure resulting from systematic examination of the task to be performed. This involves identify all the hazards and appropriate control measures balanced against an acceptable level of risk, in relation to the outcome.

SCD – Short circuiting devices A device, normally deployed by rail professionals, which isolates the traction current to a section of rail. This is normally used in conjunction with other control measures, eg request for power off.

SCRs – Station Control Rooms Sub Surface stations will be provided with control rooms that will have a range of features and information available to an incident Commander. It is normally accessible via a protected walkway.

Signal box – A 'signal box' or 'signal cabin' is a building from which railway signals and points are controlled.

Six-foot – The term used for the space between two adjacent sets of track, irrespective of the distance involved.

Sleepers – A beam made of wood, pre- or post-tensioned reinforced concrete or steel placed at regular intervals at right angles to and under the rails.

Stabling – The act of taking a train out of service and parking it in a siding without a crew.

Standard Operating Procedures (SOPs) – Standard methods or rules in which an organisation or Fire and Rescue Service operates to carry out a routine function. Usually these procedures are written in policies and procedures and all firefighters should be well versed in their content.

SCG – Strategic Co-ordination Group

T

Tactical plan – The operational plan formulated by the incident commander taking into account the objectives to be achieved balanced against identified operational hazards.

Thermal imaging camera (TIC) – Type of thermo graphic camera that renders infrared radiation as visible light, it allows firefighters to see and in some cases record temperatures of material.

Third rail – The third rail traction conducts the electrical current from the rail to the motor of the train.

Train Operating Companies TOCs – An organisation licensed to operate trains over the rail network.

TOPS – Total Operations Processing System. A National Rail system's Computer System able to identify the location and contents of trains and individual wagons and containers.

Track/Track area – The area between the ballast shoulders of a single, double, or multi-running line railway.

Traction Current – Term used for electric power supply for electric rail vehicles. Normally supplied by overhead wire or electrified rail and collected by a pantograph on the roof of the train in the former case or by shoes attached to the bogies in the latter.

Traction Current shut down – Means the same as isolation.

Trailer car – Is any car that does not carry traction or power related equipment and is similar to a passenger car in a locomotive hauled train.

Trains stopped – No train movement on a given section of track.

Trains stopped and power off' – No train movement on a given section of track and the electrical traction current for that section is shut down.

U

'Up Line' – A rail industry term for indicating the direction of rail vehicle movement. FRS personnel should exercise care when such terminology is used and ensure the direction of rail vehicle travel is confirmed.

Urban Search and Rescue (USAR) – Specialist teams FRS teams that are equipped to deal with incidents involving the location, extrication, and initial medical stabilisation of casualties trapped in confined spaces.

Fire and Rescue Service Operational Guidance – Railway Incidents

Section 12
References/Supporting information

Fire and Rescue Service Operational Guidance – Railway Incidents

References and Bibliography

HM Government 1974. *Health and Safety at Work etc Act 1974.* London: HMSO

HM Government, 1990. *Environmental Protection Act.* London: HMSO

HM Government, 1991. *Water Resources Act 1991.* London: HMSO

HM Government, 1999. *Management of Health and Safety at Work Regulations 1999.* London: HMSO

HM Government, 2004. *Civil Contingencies Act.* CCA (Contingency Planning) Regulations 2005 London: HMSO

HM Government, 2004. *Fire and Rescue Services Act 2004.* London: HMSO

HM Government, 2007. Fire and Rescue Services (Emergencies) (England), Order 2007

HM Government, 2008. *Fire and Rescue Manual, Fire Service Operations Vol. 2 Incident Command.* London: TSO

HM Government 2005 The Railways (Accident Investigation and Reporting) Regulations 2005 (RAIR)

HM Government 2009 The Carriage of Dangerous Goods and Use of Transportable Pressure Equipment Regulations 2009

HM Government 1977 Conveyance of Military Explosives Regulations 1977

HM Government 2006 The Railways and Other Guided Transport Systems (Safety) Regulations 2006

DCLG, 2007. *The Fire and Rescue Services (Emergencies) (England) Order 2007.* London: HMSO

Network Rail document RT/CM/SO/P/302 Railway Safety for the Emergency Services

Network Rail document RT/CM/SO/P/303 Management of Fatalities on the Railway

Network Rail document RT/CM/SO/P/305 Emergency Services Rail Incident Protocol

Network Rail document RT/CM/SO/P/306 Dangerous Goods on the Railway

Guide to Risk Assessment tools, techniques and data- Fire Research Series 5/2009

RSSB Railway Group Standard Publication GO/RT3422

RSSB Railway Group Standard Publication GO/RT3053

RSSB Railway Group Standard Publication GM/RT2456

HM Government NPIA – Guidance on Emergency Procedures 2009

RAIB/HSE/BTP – Memorandum of Understanding

Reports/Inquiries

Kings Cross (Fennel, 1987)

Ladbroke Grove (Cullen, 1999)

Channel Tunnel Fire (HSE 1996)

Baku Underground Metro 1995

Fire and Rescue Guidance Documents

Avon Fire and Rescue Service, January 2009 Fires and Incidents on the Railways A14

Cornwall Fire and Rescue Service, September 2008 Railway Incidents (OPS1/014)

Cumbria Fire and Rescue Service, Railway Incidents

Derbyshire Fire and Rescue Service, July 2008 Railway Incidents Operational Note V1.0 & SOP V1.0

Dorset Fire and Rescue Service, December 2006 Transport Systems (Rail) Sis/Swp TS 02

East Sussex Fire and Rescue Service, June 2008 Railway Incidents

Kent Fire and Rescue Service, July 2009 Incidents on Railway Property Operational Memorandum 26/094_01 V1

Hampshire Fire and Rescue Service, 2008 Railway Incidents, SO/7/6/5

Humberside Fire and Rescue Service, September 2007 Incidents Involving Rail Transport Systems SOP4.2

London Fire Brigade, January 2003, Railway Procedures Policy No 316

Merseyside Fire and Rescue Service, July 2009 Transport Systems – Rail SOP 4.2

North Yorkshire Fire and Rescue Service, February 2009 Incidents Involving Transport – Rail SOP 4.2

Oxfordshire Fire and Rescue Service, July 2008 Railway Incidents ER PROC 084 OPS PROC 043

West Midlands Fire and Rescue Service, July 2009 Railway Incidents Operational Procedure Note 13

West Sussex Fire and Rescue Service, 2005 Railway Related Incidents, Working on or near Railways SOP

West Yorkshire Fire & Rescue Service, September 2008 Incidents On Railways Operational Procedure No 45 & SOP 3/02

Wiltshire Fire & Rescue Service, August 2007 Railway Incidents Standard Operating Procedure No. 020

Web Sites

National

www.raib.gov.uk

www.rssb.co.uk

www.rail-reg.gov.uk

www.communities.gov.uk

www.opsi.gov.uk

www.nao.org.uk

www.dft.gov.uk/pgr/rail/

www.pteg.net/

www.networkrail.co.uk/

www.atoc.org/

www.fta.co.uk/

www.btp.police.uk/

Local

www.travelmetro.co.uk/

www.wirral.gov.uk/

www.blackpooltransport.com/

www.tfl.gov.uk/modalpages/2674.aspx

www.metrolink.co.uk/

www.thetram.net/

www.supertram.com/

www.nexus.org.uk/wps/wcm/connect/Nexus/Metro

www.bluebell-railway.co.uk

Section 13

Record of obsolete or superseded previous operational guidance

Fire and Rescue Service Operational Guidance – Railway Incidents

The table below lists the guidance relating to railway incidents issued by Her Majesty's Government that is now deemed to be obsolete or, is superseded by this Operational Guidance document.

The following abbreviations are used in the table:

- MoF   Manual of Firemanship
- DCOL   Dear Chief Officer Letter
- FSC   Fire Service Circular

Type of guidance	Document title
DCOL 700/1964	Fire protection of diesel locomotives
DCOL 706/1964	Fire protection of diesel locomotives
DCOL 9/1992	Locomotive silencer fires
MoF	Manual of firemanship part 6c, Fires involving explosives carriage of explosives by rail
DCOL (1981/3)	Item 1. Safety procedures at fires and other incidents on railway property
DCOL (1984/3)	Item E. Procedures for incidents on railway property
MoF	Manual of firemanship Book 4, Incidents involving Aircraft, Shipping and Railways – Part 3
DCOL (1993/2)	Item 7. British Rail – emergency planning and contact arrangements with emergency services
DCOL (1993/2)	Item 8. British Rail – emergency cutting – aluminium rolling stock
DCOL (1991/10)	Item 1. Incidents on railway property
DCOL (1990/6)	Item 1. Incidents on railway property
DCOL (1989/8)	Item 4. Polychlorinated biphenyls (PCBs) – British Rail
DCOL (1988/5)	Item A. Incidents on railway property
DCOL (1987/10)	Item A. Incidents on railway property
FSC (2004/40)	40/2004 – Flooding on or near the Railway Infrastructure
FSC (2004/20)	20/2004 – Carriage of Dangerous Goods
DCOL (2004/2)	Item B train glazing
DCOL (1999/8)	Item F. Incidents involving railways
DCOL (1996/12)	Item F. New regulations on the transport of dangerous goods by road and rail

Notes

Notes